Lecture Notes in Computer Scie

Commenced Publication in 1973
Founding and Former Series Editors:
Gerhard Goos, Juris Hartmanis, and Jan van Leeuwen

Editorial Board

María J. Blesa   Christian Blum
Carlos Cotta   Antonio J. Fernández
José E. Gallardo   Andrea Roli
Michael Sampels (Eds.)

Hybrid Metaheuristics

5th International Workshop, HM 2008
Málaga, Spain, October 8-9, 2008
Proceedings

 Springer

Volume Editors

María J. Blesa
Universitat Politècnica de Catalunya, Barcelona, Spain
E-mail: mjblesa@lsi.upc.edu

Christian Blum
Universitat Politècnica de Catalunya, Barcelona, Spain
E-mail: cblum@lsi.upc.edu

Carlos Cotta
Universidad de Málaga, Málaga, Spain
E-mail: ccottap@lcc.uma.es

Antonio J. Fernández
Universidad de Málaga, Málaga, Spain
E-mail: afdez@lcc.uma.es

José E. Gallardo
Universidad de Málaga, Málaga, Spain
E-mail: pepeg@lcc.uma.es

Andrea Roli
DEIS, Università di Bologna, Cesena, Italy
E-mail: andrea.roli@unibo.it

Michael Sampels
Université Libre de Bruxelles, Bruxelles, Belgium
E-mail: msampels@ulb.ac.be

Library of Congress Control Number: 2008936281

CR Subject Classification (1998): I.2.8, F.2, G.1.6, F.1, G.2

LNCS Sublibrary: SL 1 – Theoretical Computer Science and General Issues

ISSN   0302-9743
ISBN 978-3-540-88438-5 Springer Berlin Heidelberg New York

Springer is a part of Springer Science+Business Media

springer.com

© Springer-Verlag Berlin Heidelberg 2008

Typesetting: Camera-ready by author, data conversion by Scientific Publishing Services, Chennai, India
Printed on acid-free paper   SPIN: 12533855   06/3180   5 4 3 2 1 0

Preface

"You see, I have a lot of special knowledge which I apply to the problem, and which facilitates matters wonderfully," says Sherlock Holmes to Dr. Watson in *A Study in Scarlet* by Arthur Conan Doyle. The *knowledge* exploited to tackle difficult problems is probably the main theme of the papers selected for this fifth edition of the International Workshop on Hybrid Metaheuristics. Indeed, in most of the papers a specific combination of metaheuristics and other solving techniques is presented for tackling a particular relevant constrained optimization problem, such as fiber optic networks, timetabling and freight train scheduling problems. The quest for solvers which can successfully and efficiently handle *relevant problems* is the main motivation for research in metaheuristics: it is important to keep this in mind so as to clearly state our research goals and methodology. The question arises as to what is the definition of *relevant problems* and a possible answer is that any useful and even just interesting or funny problem can be considered as scientifically relevant.

The research goal of solving relevant problems does not require practitioners to assemble some software code and, with a little faith in alchemy, hope that the outcome is a reasonably good solution. On the contrary, this research must be grounded on a scientific method and on technological skills. That is why it is so important to support the assessment of an algorithm's performance with a sound methodology. This requires studying theoretical models for describing properties of the hybrid metaheuristics, and to be open to other communities and to compare our achievements with theirs.

We would like this to be the view of the participants of the International Workshop on Hybrid Metaheuristics combined with tangible improvements in producing scientifically grounded results. The selection of papers should be useful to researchers both in finding new ideas and for implementing efficient solutions.

As in previous editions of this workshop, special care was taken in the reviewing process: out of 33 submissions received, 14 papers were selected on the basis of the reviews by the Program Committee members and evaluations by the Program Chairs. The review process was systematic and intended for providing authors with constructive suggestions for improvements. Our special thanks go to the Program Committee members for their devoted efforts.

An agenda for future research in hybrid metaheuristics could focus on three objectives: (a) the field should become more rigorous, (b) results need to be compared with those produced by other techniques, (c) new application areas should be explored. Rigour is important to earn the acceptance of our colleagues in related communities. The comparison with related fields is valuable in assessing the effectiveness of hybrid techniques and in developing improved hybrid algorithms. New application areas can provide us with exciting new problems that have both stochastic and online elements, which arise both in simulation and data mining. In other words, *"a lot of special knowledge ... which facilitates matters wonderfully."*

August 2008

María J. Blesa
Christian Blum
Carlos Cotta
Antonio J. Fernández
José E. Gallardo
Andrea Roli
Michael Sampels

Organization

General Chair

Carlos Cotta — Universidad de Málaga, Spain

Program Chairs

María J. Blesa — Universitat Politècnica de Catalunya, Barcelona, Spain

Christian Blum — Universitat Politècnica de Catalunya, Barcelona, Spain

Antonio J. Fernández — Universidad de Málaga, Spain

José E. Gallardo — Universidad de Málaga, Spain

Andrea Roli — Università di Bologna, Italy

Michael Sampels — Université Libre de Bruxelles, Belgium

Program Committee

Chris Beck — University of Toronto, Canada

Mauro Birattari — Université Libre de Bruxelles, Belgium

Jürgen Branke — Universität Karlsruhe, Germany

Marco Chiarandini — Syddansk Universitet, Denmark

Luca Di Gaspero — Università di Udine, Italy

Marco Dorigo — Université Libre de Bruxelles, Belgium

Karl Dörner — Technische Universität Wien, Austria

Michael Emmerich — Universiteit Leiden, The Netherlands

Antonio J. Fernández — Universidad de Málaga, Spain

José E. Gallardo — Universidad de Málaga, Spain

William Havens — Simon Fraser University, Canada

Thomas Jansen — Technische Universität Dortmund, Germany

Joshua Knowles — University of Manchester, UK

Andrea Lodi — Università di Bologna, Italy

Daniel Merkle — Syddansk Universitet, Denmark

Bernd Meyer — Monash University, Australia

Martin Middendorf — Universität Leipzig, Germany

J. Marcos Moreno — Universidad de La Laguna, Spain

José A. Moreno — Universidad de La Laguna, Spain

Boris Naujoks — Technische Universität Dortmund, Germany

David Pelta — Universidad de Granada, Spain

Steven Prestwich — 4C, Cork, Ireland

Christian Prins — Université de Technologie de Troyes, France

Günther Raidl	Technische Universität Wien, Austria
Günter Rudolph	Technische Universität Dortmund, Germany
Andrea Schaerf	Università di Udine, Italy
Marc Sevaux	Université Européenne de Bretagne, France
Thomas Stützle	Université Libre de Bruxelles, Belgium

Additional Referees

Valentina Cacchiani, Maria Kandyba, Marco Lübbecke

Table of Contents

An Evolutionary ILS-Perturbation Technique

Manuel Lozano[1] and C. García-Martínez[2]

[1] Department of Computer Science and Artificial Intelligence, University of Granada,
Granada 18071, Spain
[2] Department of Computing and Numerical Analysis, University of Córdoba,
Córdoba 14071, Spain

Abstract. This contribution proposes a new perturbation technique for the iterated local search metaheuristic, which consists in a micro evolutionary algorithm that effectively explores the neighborhood of the solution that should undergo the perturbation operator. Its main idea is to play the same role as the standard ILS-perturbation operator, but more satisfactorily. A new model of integrative hybrid metaheuristic is obtained by incorporating the proposed perturbation approach into the iterated local search algorithm, because the evolutionary algorithm becomes a subordinate component of iterated local search. The benefits of the proposal in comparison to other iterated local search algorithms proposed in the literature to deal with binary optimization problems are experimentally shown.

1 Introduction

In the last few years, a new family of search and optimization algorithms have arisen based on extending basic heuristic methods by including them into an iterative framework augmenting their exploration capabilities. This group of advanced approximate algorithms has received the name *metaheuristics* (MHs) [5] and an overview of various existing methods is found in [1]. MHs have proven to be highly useful for approximately solving difficult optimization problems in practice because they may obtain good solutions in a reduced amount of time. Simulated annealing, tabu search, evolutionary algorithms (EAs), ant colony optimization, estimation of distribution algorithms, scatter search, path relinking, GRASP, multi-start and iterated local search, guided local search, and variable neighborhood search are, among others, often listed as examples of classical MHs, and they have individual historical backgrounds and follow different paradigms and philosophies.

Over the last years, a large number of search algorithms were reported that do not purely follow the concepts of one single classical MH, but they attempt to obtain the best from a set of MHs (and even other kinds of optimization methods) that perform together and complement each other to produce a profitable synergy from their combination. These approaches are commonly referred to as *hybrid* MHs [19]. Hybrid MHs may be categorized into two different classes attending on their control strategy [16]: 1) *collaborative* hybrid MHs, which are

M.J. Blesa et al. (Eds.): HM 2008, LNCS 5296, pp. 1–15, 2008.

based on the exchange of information between different MHs (and possibly other optimization techniques) running sequentially or in parallel, and 2) *integrative* hybrid MHs, where one search algorithm is considered a subordinate, embedded component of another algorithm.

Iterated local search (ILS) [9,13] belongs to the group of MHs that extend classical local search (LS) methods by adding diversification capabilities. The essential idea of ILS is to perform a biased, randomized walk in the space of locally optimal solutions instead of sampling the space of all possible candidate solutions. This walk is built by iteratively applying first a *perturbation* to a locally optimal solution, then applying a LS algorithm, and finally using an acceptance criterion which determines to which locally optimal solution the next perturbation is applied. Despite its simplicity, it is at the basis of several state-of-the-art algorithms for real-world problems [9].

An important aspect of ILS is the mechanism to perform perturbations. This may be a random mechanism or it may be produced by a semi-deterministic method (e.g., a LS different from the one used in the main algorithm [1]). In this respect, we should point out that, nowadays, an attractive line of research on ILS concerns the use of different EA principles to build innovative perturbation models [11,12,21,24].

With the aim of providing additional results and insights on this line of research, in this contribution, we design a *customized* EA playing the same role as the standard ILS-perturbation operator, but more satisfactorily. In particular, we present an *evolutionary ILS-perturbation mechanism* that involves a micro EA that explores the neighborhood of the solution that should undergo the perturbation operator. By incorporating the proposed perturbation approach into ILS, we transform this classical MH into an integrative hybrid MH, because one of its components is another MH (an EA).

The remainder of this paper is organized as follows. In Section 2, we give an overview of ILS. In Section 3, we propose the evolutionary perturbation technique and describe the way it is integrated in ILS. In Section 4, we show the benefits of the proposal in comparison to other ILS algorithms proposed in the literature to deal with binary optimization problems. Finally, in Section 5, we provide the main conclusions of this work and examine future research lines.

2   Iterated Local Search

A high level description of ILS as it is described in [13] is given in Figure 1. The algorithm starts by applying LS to an initial solution and iterates a procedure where a perturbation is applied to the current solution S^* in order to move it away from its local neighborhood; the solution so obtained is then considered as initial point for a new LS processing, resulting in another locally optimal solution S_{LS}. Then, a decision is made between S^* and S_{LS} to decide from which solution the next iteration continues.

The perturbation operator is a key aspect to consider, because it allows ILS to reach a new solution from the set of local optima by escaping from basis of

```
Iterated Local Search

    1. S₀ ← GenerateInitialSolution()
    2. S* ← LocalSearch(S₀)
    3. while (termination conditions not met) do
        4. S_P ← Perturbation(σ_p, S*, history)
        5. S_LS ← LocalSearch(S_P)
        6. S* ← AcceptanceCriterion(S*, S_LS, history)
    7. return S*
```

Fig. 1. Pseudocode algorithm for ILS

attraction of the previous local optimum. The perturbation is usually nondeterministic in order to avoid cycling. For example, for the case of binary problems, the perturbation operator flips the bits with a fixed probability. Its most important characteristic is the perturbation strength (σ_p), roughly defined as the amount of changes made on the current solution. The perturbation strength should be large enough such that the LS does not return to the same local optimum in the next iteration. However, it should not be too large; otherwise the search characteristics will resemble those of a multi-start LS algorithm.

An important aspect in the perturbation and the acceptance criterion is to introduce a bias between intensification and diversification of the search. Intensification can be reached by applying the perturbation always to the best solution found and using small perturbations. On the other hand, diversification is achieved by accepting every new solution S^* and applying large perturbations. Then, the perturbation operator arises as one of the most determinant component of ILS. In addition, it is a key aspect to consider in the design of ILS; as claimed by [13]: *"A good perturbation transforms one excellent solution into an excellent starting point for a local search"*.

The mechanism to perform perturbations may be a random mechanism (as was aforementioned), or it may be produced by a semi-deterministic method (e.g., a LS different from the one used in the main algorithm [1]). In this respect, we should point out that different EA principles have been used to build new perturbation models. Examples are:

- *Population-based ILS* (PILS). Thierens [21] proposes a MH that combines the power of ILS with the principle of extracting useful information about the search space by keeping a population of solutions. In addition to ILS, PILS also keeps a small population of neighboring solutions and restricts the perturbation of ILS to the subspace where the current solution and a population member disagree, thus preserving their common substructure. The key assumption of the PILS algorithm is that local optimal solutions possess common substructures that can be exploited to increase the efficiency of ILS.
- *Genetic ILS* (GILS). Katayama et al. [11,12] introduced a new perturbation mechanism for an ILS instance developed for the traveling salesman problem,

which was called genetic ILS. This perturbation approach uses a crossover operator specifically designed to deal with this problem. In each iteration, GILS perturbs the best found solution, S_{best}, generating S_P. Then, it applies the crossover operator to S_{best} and S_P, producing the final solution that will be refined by the LS operator.

- *ILS with guided mutation* (ILS/GM). Zhang et al. [24] used the *guided mutation operator* [23] as perturbation operator in ILS for the quadratic assignment problem. Guided mutation uses the idea of estimation of distribution algorithms to improve conventional mutation operators. It provides a mechanism for combining global statistical information about the search space and the position information of a good solution found during the previous search for generating new trial solution.

3   Evolutionary ILS-Perturbation Technique

As was clearly stated above, a promising research line that may be followed to improve ILS performance involves the utilization of different EA principles to design the perturbation mechanism. In this section, we present an evolutionary ILS-perturbation technique, which is based on the EA model called CHC [2] (Appendix A). It will be denominated μCHC. Our main idea is to build a new ILS model, called ILS-μCHC, which follows pseudo-code in Figure 1 replacing Step 4 by:

4. $S_P \leftarrow \mu\text{CHC}(\sigma_p, S^*)$

We have conceived μCHC to be an effective *explorer* in the neighborhood of S^*. At the beginning of this algorithm, S^* is used to create its initial population. Then, it is performed throughout a predetermined number of fitness function evaluations. The best reached individual is then considered as starting point for the next LS process (Figure 2).

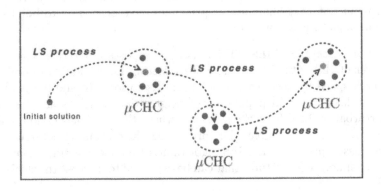

Fig. 2. ILS with evolutionary perturbation technique

We have chosen the CHC algorithm as basis to build our evolutionary ILS-perturbation method because it suitably combines powerful diversification mechanisms with an elitist selection strategy. The filtering of high diversity by means of high selective pressure favors the creation of *useful diversity*; many dissimilar solutions are produced during the run and only the best ones are conserved in the population, allowing diverse and promising solutions to be maintained. From our point of view, this behavior is desirable for an EA assuming the work of a perturbation operator

Next, we detail the main adaptations made on the original formulation of CHC to obtain our evolutionary ILS-perturbation technique:

1. *Population size.* μCHC manages a population with few individuals ($N = 5$), and thus, it may be seen as micro EA. In standard ILS models, the number of fitness function evaluations required by the perturbation mechanism is very low as compared with the one for the LS method. With the aim of preserving, as far as possible, the essence of ILS, we have considered an EA with a low sized population; for being able to work adequately under the requirement of spending reduced number of evaluations.

2. *Number of evaluations.* In particular, we have limited this number through the following strategy: the number of evaluations assigned to μCHC for a particular invocation will be a fixed proportion, p_{evals}, of the number of evaluations consumed by the previously performed LS method. It is worth noting that p_{evals} should be set to a low value.

3. *Initial population.* Every individual in the initial μCHC population is generated by performing standard perturbation on the current solution, S^*, using the perturbation strength σ_p (which becomes a parameter associated with μCHC).

4. *Cataclysmic mutation.* It fills the population with individuals created by the same way as initial population is built (by perturbing S^*) and preserves the best performing individual found in the previous evolution. After applying cataclysmic mutation, the difference threshold (Appendix A) is set to: $\sigma_p * (1 - \sigma_p) * L$ (i.e., we have considered that $\sigma_p = p_{cm}$).

5. *Mating with S^*.* Finally, we should highlight that μCHC incorporates the appealing principle in GILS [12] of recombining S^* with another solution. In addition to the typical recombination phase of CHC, our algorithm always mates S^* with an individual in the population (selected at random) and, if they are finally crossed over, attending on the incest prevention mechanism (Appendix A), the resulting offspring will be introduced into the offspring population of μCHC.

Finally, we may highlight that our proposal gathers together the idea in GILS of using a crossover operator with the one of PILS of managing a population of solutions. In this way, it attempts to combine the best of these methods.

4 Experiments

We have carried out experiments on a test suite composed by 19 binary optimization problems (Appendix B), in order to study the behavior of the ILS model based on the evolutionary perturbation mechanism presented in the previous section. Firstly, we describe the experimental setup used (Section 4.1), then, we analyze the results obtained from different experimental studies carried out with ILS-μCHC. In particular, our aim is: 1) to ascertain whether the innovative design of μCHC is suitable to allow this algorithm to outperform other ILS models with contemporary perturbation methods of the literature (Section 4.2) and 2) to compare its results with the ones for other ILS instances that were built with the specific objective of enhancing diversification (Section 4.3). The results of all executed ILS algorithms may be found in Appendix D.

4.1 Experimental Setup

In this section, we describe the basic scheme of the ILS algorithms (Figure 1) compared in our experiments. They were specifically implemented to tackle optimization problems in a fixed-length binary search space:

- *LS procedure.* It is the *first-improvement hill-climbing algorithm*, which consists in having one individual and keep mutating each gene, one at a time, in a predefined random sequence, until the resulting individual is fitter than the original. In that case, the new individual replaces the original and the procedure is repeated until no improvement can be made further.
- *Initial solution.* It is a fixed-length binary string generated at random.
- *Acceptation criterion.* We have used the requirement that new solutions should have a better (or at least equal) fitness value than the current solution.

All the algorithms were executed 50 times (initial solutions were the same for the corresponding runs for all the ILS instances), each one with a maximum of 100,000 fitness function evaluations. We have used the *Wilcoxon matched-pairs signed-ranks* test to compare the results of our proposal with the ones of other ILS approaches (see Appendix C).

4.2 Comparison of μCHC with other Perturbation Methods

The main aim of this section is to compare μCHC with other contemporary perturbation techniques of the literature. In order to do this, we have implemented several ILS algorithms that follow the basic scheme described in the previous section and are distinguished uniquely by the perturbation operator:

- ILS with *standard binary-perturbation* (SILS). This perturbation method flips the bits with a fixed probability, the perturbation strength, σ_p.
- PILS [21]. The values for the parameters associated with the perturbation operator in this algorithm are $N = 5$, $P_{ratio} = 0.5$, and $P_{maskmut} = 0.25$.

- GILS [12]. We have considered *uniform* crossover for implementing the perturbation strategy for this algorithm.
- ILS/GM [24]. We have implemented the guided mutation operator proposed in [23] to manipulate binary-coded chromosomes (β was set to 0.005). We used guidelines in [24] to adapt this operator as perturbation mechanism for ILS.

μCHC uses standard binary-perturbation for generating initial population and populations after applying cataclysmic mutation. In addition, it assumes $p_{evals} = 0.25$, i.e., at each invocation, it consumes the 25% of number of evaluations utilized by the previous processing of LS procedure.

Since all the implemented ILS algorithms are distinguished uniquely by the perturbation policy, we may determine the significance of our newly proposed method. In order to make the comparison, firstly, we investigate the influence of σ_p on the performance of these ILS algorithms. In particular, we analyze the behavior of these algorithms when different values for this parameter are considered (σ_p =0.1, 0.25, 0.5, and 0.75).

For each ILS algorithm, Figure 3 shows the average ranking obtained by its instances with different σ_p values when compared among them. This measure is obtained by computing, for each problem, the ranking r_j of the observed results for instance j assigning to the best of them the ranking 1, and to the worst the ranking k (k is the number of instances). Then, an average measure is obtained from the rankings of this instance for all test problems. For example, if a certain instance achieves rankings 1, 3, 1, 4, and 2, on five test functions, the average ranking is $\frac{1+3+1+4+2}{5} = \frac{11}{5}$. Clearly, the lower a column is, the better its associated ILS instance is.

An important remark from Figure 3 is that the best ranked instance of ILS-μCHC uses $\sigma_p = 0.25$ while ones for the other ILS algorithms employ $\sigma_p = 0.1$. This indicates that ILS-μCHC achieves its best behavior by working in more extensive neighborhoods than other ILS algorithms do, i.e., its performance becomes better by processing higher diversification levels than its competitors. Next, we investigate whether this mode of operating allows it to obtain better results. Then, we have undertaken a comparative analysis between ILS-μCHC with $\sigma_p = 0.25$ and each one of the other ILS algorithms (with all σ_p values) by means of the Wilcoxon's test (Appendix C). Table 1 summarizes the results of this procedure, where the values of R+ (associated to ILS-μCHC) and R- of the test are specified together with the critical values. Last column indicates whether our algorithm performs statistically equivalent to the other algorithm (the null hypothesis of equality of means is accepted) or there exist significant differences between them (the null hypothesis of equality of means is rejected).

From Table 1, we clearly notice that ILS-μCHC obtained improvements with regards to the other algorithms, which are statistically significant (because all R-values are lower than both R+ ones and critical values). These initial experiments suggest that our evolutionary ILS-perturbation technique may really enhance the operation of ILS and, thus, it becomes prospective for effectively exploring the neighborhood of S^*.

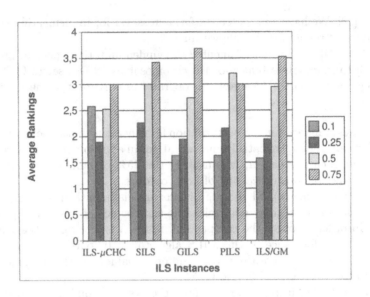

Fig. 3. Average rankings obtained by ILS instances with different σ_p values

Table 1. ILS-μCHC ($\sigma_p = 0.25$) versus ILS algorithms with other perturbation models (Wilcoxon's test with p-value $= 0.05$)

ILS-μCHC vs.	$R+$	$R-$	Critical value	Sig. differences?
SILS ($\sigma_p = 0.1$)	177.0	13.0	46	yes
SILS ($\sigma_p = 0.25$)	186.0	4.0	46	yes
SILS ($\sigma_p = 0.5$)	183.0	7.0	46	yes
SILS ($\sigma_p = 0.75$)	171.0	19.0	46	yes
PILS ($\sigma_p = 0.1$)	180.0	10.0	46	yes
PILS ($\sigma_p = 0.25$)	186.0	4.0	46	yes
PILS ($\sigma_p = 0.5$)	173.0	17.0	46	yes
PILS ($\sigma_p = 0.75$)	164.0	26.0	46	yes
GILS ($\sigma_p = 0.1$)	170.0	20.0	46	yes
GILS ($\sigma_p = 0.25$)	182.0	8.0	46	yes
GILS ($\sigma_p = 0.5$)	187.0	3.0	46	yes
GILS ($\sigma_p = 0.75$)	186.0	4.0	46	yes
ILS/GM ($\sigma_p = 0.1$)	167.0	23.0	46	yes
ILS/GM ($\sigma_p = 0.25$)	185.0	5.0	46	yes
ILS/GM ($\sigma_p = 0.5$)	187.0	3.0	46	yes
ILS/GM ($\sigma_p = 0.75$)	185.0	5.0	46	yes

4.3 Comparison with ILS Approaches That Use Diversification Techniques

In this section, we compare ILS-μCHC with other ILS models built by enhancing diversification properties of standard ILS:

- *ILS with random walk acceptance* (ILS-RW). The acceptance criterion can roughly be used to control the balance between intensification and diversification for ILS search [13]. A simple way to illustrate this is to consider a *Markovian* acceptance criterion. A very strong intensification is achieved if only better solutions are accepted. At the opposite extreme is the random walk acceptance criterion (denoted by RW) which always applies the perturbation to the most recently visited local optimum, irrespective of its cost. This criterion clearly favors diversification over intensification, because it promotes a stochastic search in the space of local optima.

- *Collaborative ILS* (CILS). Another convenient way of empowering ILS exploration involves the idea of replacing a single ILS run by a population of ILS runs that interact each others in some way [18] (we call this ILS model *collaborative* ILS). The aim of this strategy is to avoid a stagnation behavior by, in some sense, delaying the decision on which solution one has to concentrate to find the highest solution quality; by the use of a population of ILS runs, the algorithm is not forced to concentrate the search only around the best solution found as done in single ILS runs.

We have implemented a variant of collaborative ILS, called replace-worst, that starts with λ solutions each of which follows a standard ILS algorithm, except that every n_{It} iterations a copy of the current best solution replaces the worst solution in the population. We have considered $\lambda = 20$ and tried two different situations with regards to the communication strategy among ILS runs: 1) without communication (i.e., multiple independent trials of an ILS algorithm) and 2) with $n_{It} = 3$. They are denoted as CILS-wc and CILS-3, respectively.

The performance comparison among ILS-μCHC and each one of these algorithms was carried out by means of the Wilcoxon's test. Table 2 has the results for p-value = 0.05.

Results of the Wilcoxon's test in Table 2 advise us that our algorithm consistently outperforms all ILS algorithms based on strategies to enhance diversification. In particular, an interesting remark from improvement on CILS is that it becomes more fruitful executing only one ILS instance with a perturbation operator that focuses diversification by managing a set of solutions (such as ILS-μCHC does) than favoring diversification by keeping multiple collaborative ILS runs that employ standard perturbation operator.

Table 2. ILS-μCHC ($\sigma_p = 0.25$) versus ILS algorithms with enhanced diversification (Wilcoxon's test with p-value = 0.05)

ILS-μCHC vs.	$R+$	$R-$	Critical value	Sig. differences?
ILS-RW ($\sigma_p = 0.1$)	190.0	0.0	46	yes
ILS-RW ($\sigma_p = 0.25$)	190.0	0.0	46	yes
ILS-RW ($\sigma_p = 0.5$)	185.0	5.0	46	yes
ILS-RW ($\sigma_p = 0.75$)	163.0	27.0	46	yes
CILS-wc ($\sigma_p = 0.1$)	182.0	8.0	46	yes
CILS-wc ($\sigma_p = 0.25$)	182.0	8.0	46	yes
CILS-wc ($\sigma_p = 0.5$)	179.0	11.0	46	yes
CILS-wc ($\sigma_p = 0.75$)	163.0	27.0	46	yes
CILS-3 ($\sigma_p = 0.1$)	181.0	9.0	46	yes
CILS-3 ($\sigma_p = 0.25$)	180.0	10.0	46	yes
CILS-3 ($\sigma_p = 0.5$)	179.0	11.0	46	yes
CILS-3 ($\sigma_p = 0.75$)	163.0	27.0	46	yes

These results and ones obtained in previous sections allow us to conclude that our evolutionary perturbation technique may really enhance the operation of the ILS algorithm; in fact, the ILS instance performing this perturbation technique resulted very competitive with the state-of-the-art on this well-known MH for binary optimization problems.

5 Conclusions

In this paper, we presented an evolutionary perturbation technique that may replace, without great difficulties, the standard perturbation operator of ILS, developing their work more effectively and with a relatively low computational cost. An outstanding remark is that the resulting ILS model becomes an integrative hybrid MH. The proposal has turned out to be very competitive with state-of-the-art ILS models for binary optimization problems. Therefore, the research line focused in this paper is indeed worth of further studies. We are currently extending our investigation to build evolutionary ILS-perturbation techniques being well-suited to deal with real-world problems, such as the traveling salesman problem, the quadratic assignment problem, the p-median problem, etc.

Acknowledgment. This work was supported by Project TIN2005-08386-C05-01.

References

1. Blum, C., Roli, A.: Metaheuristics in combinatorial optimization: overview and conceptual comparison. ACM Comput. Surv. 35(3), 268–308 (2003)
2. Eshelman, L.J.: The CHC adaptive search algorithm: how to have safe search when engaging in non-traditional genetic recombination. In: Rawlins, G. (ed.) Foundations of Genetic Algorithms 1, pp. 265–283. Morgan Kaufmann, California (1991)
3. Forrest, S., Mitchell, M.: Relative building block fitness and the building block hypothesis. In: Whitley, L.D. (ed.) Foundations of Genetic Algorithms 2, pp. 109–126. Morgan Kaufmann, California (1993)
4. García-Martínez, C., Lozano, M., Herrera, F., Molina, D., Sánchez, A.M.: Global and local real-coded genetic algorithms based on parent-centric crossover operators. Eur. J. Oper. Res. 185, 1088–1113 (2008)
5. Glover, F., Kochenberger, G. (eds.): Handbook of metaheuristics. Kluwer Academic Publishers, Massachusetts (2003)
6. Goldberg, D.E., Korb, B., Deb, K.: Messy genetic algorithms: motivation, analysis, and first results. Complex Syst. 3, 493–530 (1989)
7. Goldberg, D.E., Deb, K., Horn, J.: Massively multimodality, deception, and genetic algorithms. In: Männer, R., Manderick, B. (eds.) Proc. of the Int. Conf. on Parallel Problem Solving from Nature, pp. 37–46. North-Holland Pub. Co., Amsterdam (1992)
8. Heitktter, J.: SAC-94 Suite of 0/1-multiple-knapsack problem instances (2001), http://elib.zib.de/pub/Packages/mp-testdata/ip/sac94-suite

9. Hoos, H.H., Stützle, T.: Stochastic local search:–foundations and applications. Morgan Kaufmann Publishers, California (2004)
10. Hu, J., Goodman, E., Seo, K., Fan, Z., Rosenberg, R.: The hierarchical fair competition (HFC) framework for sustainable evolutionary algorithms. Evol. Comput. 13(2), 241–277 (2005)
11. Katayama, K., Narihisa, H.: Iterated local search approach using genetic transformation to the travelling salesman problem. In: Banzhaf, W., Daida, J., Eiben, A.E., Garzon, M.H., Honovar, V., Jakiela, M., Smith, R.E. (eds.) Proc. of the Genetic and Evolutionary Computation Conf., pp. 321–328. Morgan Kaufmann, California (1999)
12. Katayama, K., Narihisa, H.: A new iterated local search algorithm using genetic crossover for the travelling salesman problem. In: Proc. of the ACM Symposium on Applied Computing, pp. 302–306 (1999)
13. Lourenço, H.R., Martin, O.C., Stützle, T.: Iterated local search. In: Glover, F., Kochenberger, G. (eds.) Handbook of Metaheuristics, pp. 321–353. Kluwer Academic Publishers, Massachusetts (2003)
14. Pelikan, M., Goldberg, D.E., Cantú-Paz, E.: Linkage problem, distribution estimation, and bayesian networks. Evolutionary Comput. 8(3), 311–340 (2000)
15. Pelikan, M., Hartmann, A., Lin, T.K.: Parameter-less hierarchical bayesian optimization algorithm. In: Lobo, F.G., Lima, C.F., Michalewicz, Z. (eds.) Parameter Setting in Evolutionary Algorithms. SCI, vol. 54, pp. 225–239. Springer, Berlin (2007)
16. Raidl, G.R.: A unified view on hybrid metaheuristics. In: Almeida, F., Blesa Aguilera, M.J., Blum, C., Moreno Vega, J.M., Pérez Pérez, M., Roli, A., Sampels, M. (eds.) HM 2006. LNCS, vol. 4030, pp. 1–12. Springer, Heidelberg (2006)
17. Sheskin, D.J.: Handbook of parametric and nonparametric statistical procedures. CRC Press, Boca Raton (2003)
18. Stützle, T.: Iterated local search for the quadratic assignment problem. Eur. J. Oper. Res. 174, 1519–1539 (2006)
19. Talbi, E.-G.: A taxonomy of hybrid metaheuristics. J. Heuristics 8(5), 541–565 (2002)
20. Thierens, D.: Adaptive mutation rate control schemes in genetic algorithms. In: Proc. of the Congress on Evolutionary Computation, pp. 980–985. IEEE Press, Los Alamitos (2002)
21. Thierens, D.: Population-based iterated local search: restricting neighborhood search by crossover. In: Deb, K., et al. (eds.) GECCO 2004. LNCS, vol. 3103, pp. 234–245. Springer, Heidelberg (2004)
22. Zar, J.H.: Biostatistical analysis. Prentice-Hall, Englewood Cliffs (1999)
23. Zhang, Q., Sun, J., Tsang, E.P.K.: Evolutionary algorithm with the guided mutation for the maximum clique problem. IEEE Trans. Evol. Comput. 9(2), 192–200 (2005)
24. Zhang, Q., Sun, J.: Iterated local search with guided mutation. In: IEEE Congress on Evolutionary Computation, pp. 924–929. IEEE Press, Los Alamitos (2006)

A   CHC Algorithm

The key idea of the CHC algorithm [2] concerns the combination of a selection strategy with a very high selective pressure and several components inducing a strong diversity. The four main components of the algorithm are shown as follows:

- *Elitist selection.* The N members of the current population are merged with the offspring population obtained from it and the best N individuals are selected to compose the new population.
- *Half uniform crossover.* It is a highly disruptive crossover that crosses over exactly half of the non-matching alleles (the bits to be exchanged are chosen at random without replacement).
- *Incest prevention mechanism.* During the reproduction step, each member of the parent (current) population is randomly chosen without replacement and paired for mating. Before mating, the Hamming distance between the potential parents is calculated and if half this distance does not exceed a *difference threshold d*, they are not mated and no offspring coming from them is included in the offspring population. The aforementioned threshold is usually initialized to $\frac{L}{4}$ (with L being the chromosome length). If no offspring is obtained in one generation, the difference threshold is decremented by one.
- *Cataclysmic mutation.* CHC uses no mutation in the classical sense of the concept, but instead, it goes through a process of cataclysmic mutation when the population has converged. The difference threshold is considered to measure the stagnation of the search, which happens when it has dropped to zero and several generations have been run without introducing any new individual in the population. Then, the population is reinitialized by considering the best individual as the first chromosome of the new population and generating the remaining $N-1$ ones by randomly flipping a number of its bits, determined by the cataclysmic mutation rate, p_{cm} (usually $p_{cm} = 0.35$). After invoking the cataclysmic mutation, the difference threshold is reinitiated to: $p_{cm} * (1 - p_{cm}) * L$.

B   Test Suite

The test suite that we have used for different experiments consists of 19 binary-coded optimization problems. Table 3 shows their names, reference where a detailed description may be found, the length of the binary solutions (L), whether they are formulated as maximization or minimization problems, and finally, the fitness value of the global optimum. Since problems from f_{Sch} to f_{SLE} are defined on continuous domains, their variables were encoded into bit strings using binary reflected Gray coding, with 20 binary genes assigned to each variable. The dimension of the search space is 10 for f_{SLE} and 5 for the remaining continuous test functions.

C   The Wilcoxon Matched-Pairs Signed-Ranks Test

Wilcoxon's test is used for answering this question: *do two samples represent two different populations?* It is a non-parametric procedure employed in a hypothesis testing situation involving a design with two samples. It is the analogous of the paired t-test in non-parametrical statistical procedures; therefore, it is a pairwise test that aims to detect significant differences between the behavior of two algorithms.

The null hypothesis for Wilcoxon's test is $H_0 : \theta_D = 0$; in the underlying populations represented by the two samples of results, the average of the difference scores equals zero. The alternative hypothesis is $H_1 : \theta_D \neq 0$, but also can be used $H_1 : \theta_D > 0$ or $H_1 : \theta_D < 0$ as directional hypothesis.

In the following, we describe the tests computations. Let d_i be the difference between the performance scores of the two algorithms on i-th out of N functions. The differences are ranked according to their absolute values; average ranks are assigned in case of ties. Let R^+ be the sum of ranks for the functions on which the second algorithm outperformed the first, and R^- the sum of ranks for the opposite. Ranks of $d_i = 0$ are split evenly among the sums; if there is an odd number of them, one is ignored:

$$R^+ = \sum_{d_i > 0} rank(d_i) + \frac{1}{2} \sum_{d_i = 0} rank(d_i) \text{ and}$$

$$R^- = \sum_{d_i < 0} rank(d_i) + \frac{1}{2} \sum_{d_i = 0} rank(d_i)$$

Table 3. Test suite

Name	Ref.	L	Max/Min	Fit. op.
Deceptive problem (D)	[6]	120	Max	900
Massively multimodal deceptive problem (MMD)	[7]	240	Max	40
Bipolar deceptive problem (BD)	[14]	120	Max	20
Overlapping deceptive problem (OD)	[14]	150	Max	74
Trap problem (T)	[21]	180	Max	1100
Trap-5 problem (T5)	[15]	150	Max	150
Royal road problem (RR)	[3]	200	Max	200
Hierarchical if-and-only-if problem	[10]			
HIFF1		128	Max	1024
HIFF2		256	Max	2304
Zero/one multiple knapsack problem	[20]			
K1 (weing7 [8])		105	Max	1095445
K2 (weish26 [8])		90	Max	9584
Schwefel's function 2.21 (f_{Sch})	[4]	100	Min	0
Quartic noise function (f_{QN})	[4]	100	Min	0
Rotated generalized Rastrigin's function (f_{RRas})	[4]	100	Min	0
Generalized Griewank function (f_{Gri})	[4]	100	Min	0
Composed $f_{Gri} - f_{Ros}$ (f_C)	[4]	100	Min	0
Schaffer's function (f_{Scha})	[4]	100	Min	0
Expanded F10 (EF_{10})	[4]	100	Min	0
Systems of linear equations (f_{SLE})	[4]	200	Min	0

Table 4. Results of the algorithms

Algorithm (σ_P)	D	MMD	BD	OD	T	T5	RR	HIFF1	HIFF2	K1	K2	f_{Sch}	f_{QN}	f_{RRas}	f_{Gri}	f_C	f_{Scha}	EF_{10}	f_{SLE}
ILS-μCHC (0.1)	848.3	30.8	18.9	72.7	1057.9	121.8	126.7	568.1	1025.4	1093127	11771.1	1.2e-4	3.2e-4	2.2	2.2e-2	1.4e-1	8.2e-2	2.0	389.3
ILS-μCHC (0.25)	853.6	30.8	18.6	71.5	1065.8	122.8	134.4	606.2	1097.4	1093321	11757.6	1.8e-4	4.8e-4	5.2e-1	1.1e-1	1.0e-1	1.1e-1	1.8	270.9
ILS-μCHC (0.50)	853.8	29.7	18.5	70.6	1062.2	123.7	130.9	595.0	1067.2	1093070	11702.0	2.1e-4	6.3e-4	5.0e-1	1.8e-1	1.5e-1	1.7e-1	1.5	231.7
ILS-μCHC (0.75)	869.6	29.4	18.5	70.0	1057.0	129.6	129.8	548.6	915.4	1092181	11697.5	3.3e-4	6.9e-4	4.2e-1	2.0e-2	1.1e-1	1.5e-1	1.6	266.7
SILS (0.1)	850.6	28.6	18.6	71.0	999.3	121.9	88.6	512.2	845.4	1084554	11358.3	3.3e-4	6.8e-4	6.6e-1	1.6e-2	2.3e-1	8.1e-1	6.9	339.4
SILS (0.25)	853.3	28.3	18.6	69.8	984.5	123.3	53.4	440.3	784.5	1071749	11218.4	7.7e-4	2.3e-3	6.8e-1	3.2e-2	2.9e-1	2.5	7.9	475.9
SILS (0.5)	858.1	28.3	18.6	69.0	984.3	125.6	36.6	414.6	763.4	1059084	10677.0	8.3e-4	7.1e-3	1.4	3.4e-2	2.6e-2	5.0	1.3e+1	523.2
SILS (0.75)	866.3	28.3	18.6	69.2	979.9	131.1	33.4	439.0	782.2	1053160	10091.3	9.7e-4	1.3e-2	2.8	4.1e-2	2.6e-2	8.4	1.5e+1	670.3
GILS (0.1)	849.6	28.4	18.7	71.7	1005.2	122.4	77.6	553.3	913.4	1086851	11442.8	1.5e-4	5.2e-4	8.4e-1	1.2e-2	2.1e-1	1.3	5.2	335.2
GILS (0.25)	851.2	28.4	18.7	70.7	995.8	122.6	78.2	481.9	833.1	1081913	11339.9	7.7e-4	7.7e-4	7.0e-1	1.8e-2	2.0e-1	1.8	5.4	412.2
GILS (0.5)	853.1	28.4	18.6	69.8	986.5	123.4	53.4	436.2	782.1	1071898	11242.8	6.4e-4	2.3e-3	7.6e-1	2.7e-2	1.9e-2	2.5	9.4	509.1
GILS (0.75)	855.6	28.2	18.6	69.3	982.1	124.4	40.8	415.6	769.6	1064176	11054.6	6.7e-4	4.6e-3	1.1	3.2e-2	2.1e-1	3.6	1.0e+1	519.4
PILS (0.1)	850.6	28.2	18.6	70.7	1004.9	121.9	75.7	510.9	879.9	1085840	11418.8	1.4e-4	4.9e-4	6.2e-1	1.2e-2	3.0e-2	6.6e-1	7.6	354.7
PILS (0.25)	851.9	28.3	18.6	70.0	998.8	123.1	61.1	459.6	822.8	1083017	11392.9	2.9e-4	7.0e-4	7.2e-1	1.6e-2	2.4e-1	7.0e-1	8.0	425.0
PILS (0.5)	857.0	28.3	18.6	69.8	992.0	125.4	57.0	449.5	811.0	1079100	11355.9	2.9e-4	1.2e-3	9.2e-1	1.6e-2	2.4e-1	2.0	9.4	449.6
PILS (0.75)	866.0	28.3	18.6	70.0	995.0	130.8	57.9	459.2	829.0	1077558	11352.9	3.4e-4	1.7e-3	1.1	1.6e-2	2.8e-1	2.3	1.1e+1	459.7
ILS/GM (0.1)	848.6	28.5	18.7	71.9	1010.2	121.7	81.1	542.1	905.4	1086622	11459.9	1.6e-4	4.5e-4	7.4e-1	1.1e-2	2.3e-1	1.0	6.0	336.6
ILS/GM (0.25)	851.7	28.3	18.7	69.9	997.3	122.2	81.1	494.3	829.4	1083295	11368.4	3.5e-4	7.9e-4	5.4e-1	1.8e-2	2.3e-1	1.0	7.7	437.5
ILS/GM (0.5)	853.2	28.3	18.6	69.7	987.8	123.2	52.5	445.2	778.8	1073329	11244.4	7.1e-4	1.9e-3	6.2e-1	2.2e-2	2.6e-1	2.4	9.2	442.7
ILS/GM (0.75)	854.8	28.4	18.6	69.4	985.3	124.4	40.0	419.1	767.3	1066857	11102.6	8.9e-4	4.6e-3	3.4	3.1e-2	2.3e-1	3.3	1.1e+1	558.6
ILS-REx (0.1)	858.2	30.1	18.9	71.5	1037.6	123.3	112.2	573.0	960.8	1089066	11618.3	2.6e-4	3.1e-4	3.4	1.8e-2	2.3e-1	1.3e-1	4.6	329.2
ILS-REx (0.25)	857.0	28.9	18.5	69.2	1005.6	126.2	68.5	471.6	830.5	1077038	11258.6	7.0e-4	8.6e-4	1.6e-1	2.1e-2	2.1e-2	3.3e-1	2.7	288.9
ILS-REx (0.50)	860.3	28.7	18.5	68.5	990.1	127.3	43.5	427.4	784.5	1064490	10816.5	8.1e-4	5.8e-3	1.4	4.7e-2	3.6e-1	2.9	7.4	459.7
ILS-REx (0.75)	868.7	29.0	18.6	69.1	987.0	132.3	42.4	469.2	833.4	1052373	10205.9	1.0e-3	1.4e-2	2.9	5.2e-2	3.3e-1	8.2	1.3e+1	606.2
ILS-μGA (0.1)	851.8	30.8	18.9	72.5	1049.7	122.2	124.6	604.7	1038.2	1091402	11694.7	1.2e-4	3.3e-4	2.5	1.5e-2	2.0e-2	8.2e-2	2.2	304.8
ILS-μGA (0.25)	857.3	29.1	18.6	69.4	1020.2	125.7	77.6	495.0	873.9	1081448	11396.3	5.5e-4	9.4e-4	2.0	1.2e-2	1.4e-2	2.2e-1	1.5	173.6
ILS-μGA (0.5)	861.1	28.8	18.5	68.5	1000.0	128.0	44.0	435.4	806.0	1067168	10887.6	8.6e-4	6.0e-3	1.2	4.0e-2	2.8e-2	1.2	2.8	368.5
ILS-μGA (0.75)	872.2	29.2	18.6	69.3	990.3	135.0	41.9	491.4	878.2	1054886	10305.1	1.1e-3	3.9e-3	2.7	5.1e-2	2.8e-2	6.1	9.5	482.7
ILS-RW (0.1)	847.2	27.1	18.3	68.0	950.6	121.8	51.0	411.1	757.3	1067993	11193.3	5.0e-4	3.8e-3	1.9	3.0e-2	2.5e-1	1.6	8.6	427.4
ILS-RW (0.25)	848.4	27.9	18.5	68.5	962.8	122.2	39.4	411.0	763.6	1062459	11128.5	7.4e-4	5.6e-3	1.6	3.3e-2	3.2e-1	4.1	1.0e+1	525.3
ILS-RW (0.5)	856.5	28.3	18.6	69.1	981.3	125.3	35.5	413.4	761.0	1060107	10628.4	9.1e-4	7.4e-3	1.6	3.6e-2	3.0e-1	5.4	1.3e+1	574.9
ILS-RW (0.75)	869.3	27.8	18.5	69.9	998.8	131.8	36.5	415.9	765.6	1057666	10433.2	9.1e-4	8.3e-3	1.6	4.8e-2	3.1e-1	5.9	1.3e+1	574.6
CILS-wc (0.1)	855.0	27.9	18.5	69.5	977.2	124.7	61.1	452.9	786.1	1078104	11294.0	5.4e-4	1.7e-3	1.2	3.2e-2	8.2e-2	3.5	4.8	454.9
CILS-wc (0.25)	855.2	28.1	18.6	69.3	975.9	124.7	45.4	424.2	776.8	1067631	11191.3	7.0e-4	3.9e-3	1.2	3.3e-2	8.2e-2	3.7	4.8	454.9
CILS-wc (0.5)	858.0	28.3	18.5	69.0	982.1	125.5	36.5	413.2	767.1	1060215	10626.9	8.7e-4	7.4e-3	1.3	3.4e-2	8.2e-2	3.6	4.8	454.9
CILS-wc (0.75)	869.7	28.0	18.6	69.7	991.9	132.3	34.9	425.3	774.0	1054873	10247.6	7.4e-4	9.0e-3	1.6	3.5e-2	8.2e-2	4.0	4.8	454.9
CILS-3 (0.1)	855.0	27.8	18.5	69.3	975.9	124.7	57.4	442.8	779.8	1074807	11263.1	5.0e-4	1.9e-3	1.3	2.8e-2	8.2e-2	3.5	4.8	454.9
CILS-3 (0.25)	854.9	28.3	18.6	69.1	974.2	124.7	44.8	422.0	774.1	1068276	11174.8	7.0e-4	3.9e-3	1.1	3.1e-2	8.2e-2	3.7	4.8	454.9
CILS-3 (0.5)	857.9	28.3	18.5	69.1	982.9	125.6	36.5	414.1	767.0	1057433	10683.8	8.6e-4	7.2e-3	1.4	3.8e-2	8.2e-2	3.6	4.8	454.9
CILS-3 (0.75)	869.6	28.1	18.5	69.7	992.9	132.2	35.8	423.1	771.1	1054968	10287.4	7.4e-4	8.8e-4	1.7	3.9e-2	8.2e-2	4.0	4.8	454.9

Let T be the smallest of the sums, $T = min(R^+, R^-)$. If T is less than or equal to the value of the distribution of Wilcoxon for N degrees of freedom (Table B.12 in [22]), the null hypothesis of equality of means is rejected.

The obtaining of the p-value associated to a comparison is performed by means of the normal approximation for the Wilcoxon T statistic (Section VI, Test 18 in [17]). Furthermore, the computation of the p-value for this test is usually included in well-known statistical software packages (SPSS, SAS, R, etc.).

D   Results of the Algorithms

Table 4 outlines the average of the best fitness function found at the end of each run for all algorithms executed for the experimental studies carried out in the paper.

A Cultural Algorithm for POMDPs
from Stochastic Inventory Control

S.D. Prestwich[1], S.A. Tarim[2], R. Rossi[1], and B. Hnich[3]

[1] Cork Constraint Computation Centre, Ireland
[2] Department of Management, Hacettepe University, Turkey
[3] Faculty of Computer Science, Izmir University of Economics, Turkey
s.prestwich@cs.ucc.ie, armagan.tarim@hacettepe.edu.tr,
r.rossi@4c.ucc.ie, brahim.hnich@ieu.edu.tr

Abstract. Reinforcement Learning algorithms such as SARSA with an eligibility trace, and Evolutionary Computation methods such as genetic algorithms, are competing approaches to solving Partially Observable Markov Decision Processes (POMDPs) which occur in many fields of Artificial Intelligence. A powerful form of evolutionary algorithm that has not previously been applied to POMDPs is the cultural algorithm, in which evolving agents share knowledge in a belief space that is used to guide their evolution. We describe a cultural algorithm for POMDPs that hybridises SARSA with a noisy genetic algorithm, and inherits the latter's convergence properties. Its belief space is a common set of state-action values that are updated during genetic exploration, and conversely used to modify chromosomes. We use it to solve problems from stochastic inventory control by finding memoryless policies for nondeterministic POMDPs. Neither SARSA nor the genetic algorithm dominates the other on these problems, but the cultural algorithm outperforms the genetic algorithm, and on highly non-Markovian instances also outperforms SARSA.

1 Introduction

Reinforcement Learning and Evolutionary Computation are competing approaches to solving Partially Observable Markov Decision Processes, which occur in many fields of Artificial Intelligence. In this paper we describe a new hybrid of the two approaches, and apply it to problems in stochastic inventory control. The remainder of this section provides some necessary background information. Section 2 describes our general approach, an instantiation, and convergence results. Section 3 describes and models the problems. Section 4 presents experimental results. Section 5 concludes the paper.

1.1 POMDPs

Markov Decision Processes (MDPs) can model sequential decision-making in situations where outcomes are partly random and partly under the control of the agent. The states of an MDP possess the *Markov property*: if the current state of the MDP at time t is known, transitions to a new state at time $t + 1$ are independent of all previous states. MDPs can be solved in polynomial time (in the size of their state-space) by modelling

M.J. Blesa et al. (Eds.): HM 2008, LNCS 5296, pp. 16–28, 2008.

them as linear programs, though the order of the polynomials is large enough to make them difficult to solve in practice [14]. If the Markov property is removed then we obtain a Partially Observable Markov Decision Process (POMDP) which in general is computationally intractable. This situation arises in many applications and can be caused by partial knowledge: for example a robot must often navigate using only partial knowledge of its environment. Machine maintenance and planning under uncertainty can also be modelled as POMDPs.

Formally, a POMDP is a tuple $\langle S, A, T, R, O, \Omega \rangle$ where S is a set of states, A a set of actions, Ω a set of observations, $R : S \times A \rightarrow \Re$ a reward function, $T : S \times A \rightarrow \Pi(S)$ a transition function, and $\Pi(\cdot)$ represents the set of discrete probability distributions over a finite set. In each time period t the environment is in some state $s \in S$ and the agent takes an action $a \in A$, which causes a transition to state s' with probability $P(s'|s,a)$, yielding an immediate reward given by R and having an effect on the environment given by T. The agent's decision are based on its observations given by $O : S \times A \rightarrow \Pi(\Omega)$.

When solving a POMDP the aim is to find a *policy*: a strategy for selecting actions based on observations that maximises a function of the rewards, for example the total reward. A policy is a function that maps the agent's observation history and its current internal state to an action. A policy may also be *deterministic* or *probabilistic*: a deterministic policy consistently chooses the same action when faced with the same information, while a probabilistic policy might not. A *memoryless* (or *reactive*) policy returns an action based solely on the current observation. The problem of finding a memoryless policy for a POMDP is NP-complete and exact algorithms are very inefficient [12] but there are good inexact methods, some of which we now describe.

1.2   Reinforcement Learning Methods

Temporal difference learning algorithms such as Q-Learning [32] and SARSA [25] from Reinforcement Learning (RL) are a standard way of finding good policies. While performing Monte Carlo-like simulations they compute a *state-action value* function $Q : S \times A \rightarrow \Re$ which estimates the expected total reward for taking a given action from a given state. (Some RL algorithms compute instead a *state value* function $V : S \rightarrow \Re$.)

The SARSA algorithm is shown in Figure 1. An *episode* is a sequence of states and actions with a first and last state that occur naturally in the problem. On taking an action that leads to a new state, the value of the new state is "backed up" to the state just left (see line 8) by a process called *bootstrapping*. This propagates the effects of later actions to earlier states and is a strength of RL algorithms. (The value γ is a *discounting factor* often used for non-episodic tasks that is not relevant for our application below: we set $\gamma = 1$.) A common *behaviour policy* is ϵ-greedy action selection: with probability ϵ choose a random action, otherwise with probability $1 - \epsilon$ choose the action with highest $Q(s,a)$ value. After a number of episodes the state-action values $Q(s,a)$ are fixed and (if the algorithm converged correctly) describe an optimum policy: from each state choose the action with highest $Q(s,a)$ value. The name SARSA derives from the tuple (s, a, r, s', a').

RL algorithms have convergence proofs that rely on the Markov property but for some non-Markovian applications they still perform well, especially when augmented with an *eligibility trace* [10,16] that effectively hybridises them with a Monte Carlo

```
1       initialise the Q(s,a) arbitrarily
2       repeat for each episode
3       (   s ← initial state
4           choose action a from s using a behaviour policy
5           repeat for each step of the episode
6           (   take action a and observe r, s'
7               choose action a' from s' using a behaviour policy
8               Q(s,a) ← Q(s,a) + α [r + γQ(s',a') − Q(s,a)]
9               s ← s',  a ← a'
10          )
11      )
```

Fig. 1. The SARSA algorithm

algorithm. We will use a well-known example of such an algorithm: SARSA(λ) [25]. When the parameter λ is 0 SARSA(λ) is equivalent to SARSA, when it is 1 it is equivalent to a Monte Carlo algorithm, and with an intermediate value it is a hybrid and often gives better results than either. Setting $\lambda > 0$ boosts bootstrapping by causing values to be backed up to states before the previous one. (See [30] for a discussion of eligibility traces, their implementation, and the relationship with Monte Carlo algorithms.) There are other more complex RL algorithms (see [13] for example) and it is possible to configure SARSA(λ) differently (for example by using *softmax* action selection instead of ϵ-greedy, and different values of α for each state-action value [30]), but we take SARSA(λ) as a representative of RL approaches to solving POMPDs. (In fact it usually outperforms two versions of Q-learning with eligibility trace — see [30] page 184.)

1.3   Evolutionary Computation Methods

An alternative approach to POMDPs is the use of Evolutionary Computation (EC) algorithms such as Genetic Algorithms (GAs), which sometimes beat RL algorithms on highly non-Markovian problems [3,19]. We shall use the most obvious EC model of POMDPs, called a *table-based representation* [19]: each chromosome represents a policy, each gene a state, and each allele (gene value) an action.

The GA we shall use is based on GENITOR [33] but without the refinements of some versions, such as genetic control of the crossover probability. This is a *steady-state* GA that, at each iteration, selects two parent chromosomes, breeds a single offspring, evaluates it, and uses it to replace the least-fit member of the population. Steady-state GAs are an alternative to *generational* GAs that generate an entire generation at each iteration, which replaces the current generation. Maintaining the best chromosomes found so far is an *elitist* strategy that pays off on many problems. Parent selection is random because of the strong selection pressure imposed by replacing the least-fit member. We use standard *uniform crossover* (each offspring gene receives an allele from the corresponding gene in a randomly-chosen parent) applied with a crossover probability p_c: if it is not applied then a single parent is selected and mutated, and the resulting chromosome replaces the least-fit member of the population. Mutation is applied to a chromosome once with probability p_m, twice with probability p_m^2, three times with probability p_m^3, and so on. The population size is P and the initial population contains random alleles.

Nondeterminism in the POMDP causes noise in the GA's fitness function. To handle this noise we adopt the common approach of averaging the fitness over a number of samples S. This technique has been used many times in *Noisy Genetic Algorithms* (NGAs) [4,6,17,18]. NGAs are usually generational and [1] show that elitist algorithms (such as GENITOR) can systematically overvalue chromosomes, but such algorithms have been successful when applied to noisy problems [29]. We choose GENITOR for its simplicity.

1.4 Hybrid Methods

Several approaches can be seen as hybrids of EC and RL. *Learning Classifier Systems* [8] use EC to adapt their representation of the RL problem. They apply RL via the EC fitness function. *Population-Based Reinforcement Learning* [11] uses RL techniques to improve chromosomes, as in a memetic algorithm. The paper is an outline only, and no details are given on how RL values are used, nor are experimental results provided. *GAQ-Learning* [15] uses Q-Learning once only in a preprocessing phase, to generate $Q(s, a)$ values. A memetic algorithm is then executed using the $Q(s, a)$ values to evaluate the chromosomes. *Q-Decomposition* [26] combines several RL agents, each with its own rewards, state-action values and RL algorithm. An arbitrator combines their recommendations, maximising the sum of the rewards for each action. It is designed for distributed tasks that are not necessarily POMPDs. Global convergence is guaranteed if the RL algorithm is SARSA but not if it is Q-Learning. In [9] a GA and RL are combined to solve a robot navigation problem. The greedy policy is applied for some time (until the robot encounters difficulty); next the GA population is evaluated, and the fittest chromosome used to update the state-action values by performing several RL iterations; next a new population is generated in a standard way, except that the state-action values are used probabilistically to alter chromosomes; then the process repeats. Several other techniques are used, some specific to robotics applications, but here we consider only the RL-EC hybrid aspects.

2 A Cultural Approach to POMDPs

A powerful form of EC is the *cultural algorithm* (CA) [21], in which agents share knowledge in a *belief space* to form a consensus. (The *belief space* of a CA is distinct from the *belief space* of a POMDP, which we do not refer to in this paper.) These hybrids of EC and Machine Learning have been shown to converge more quickly than EC alone on several applications. CAs were developed as a complement to the purely genetic bias of EC. They are based on concepts used in sociology and archaeology to model cultural evolution. By pooling knowledge gained by individuals in a body of cultural knowledge, or belief space, convergence rates can sometimes be improved. A CA has an *acceptance function* that determines which individuals in the population are allowed to *adjust* the belief space. The beliefs are conversely used to *influence* the evolution of the population. See [22] for a survey of CA applications, techniques and belief spaces. They have been applied to constrained optimisation [5], multiobjective optimisation [2], scheduling [24] and robot soccer [23], but to the best of our knowledge they have not been applied to POMDPs, nor have they utilised RL.

2.1   Cultural Reinforcement Learning

We propose a new cultural hybrid of reinforcement learning and evolutionary computation for solving POMDPs called *CUltural Reinforcement Learning* (CURL). The CURL approach is straightforward and can be applied to different RL and EC algorithms. A single set of RL state-action values $Q(s, a)$ is initialised as in the RL algorithm, and the population is initialised as in the EC algorithm. The EC algorithm is then executed as usual, except that each new chromosome is altered by, and used to alter, the $Q(s, a)$, which constitute the CA belief space. On generating a new chromosome we replace, with some probability p_l, each allele by the corresponding greedy action given by the modified $Q(s, a)$ values. Setting $p_l = 0$ prevents any learning, and CURL reduces to the EC algorithm, while $p_l = 1$ always updates a gene to the corresponding $Q(s, a)$ value, and CURL reduces to SARSA(λ) without exploration. We then treat the modified chromosome as usual by the EC algorithm: typically, fitness evaluation and placement into the population. During fitness evaluation the $Q(s, a)$ are updated by bootstrapping as usual in the RL algorithm, but the policy followed is that specified by the modified chromosome. Thus in CURL, as in several other CAs [22], *all* chromosomes are allowed to adjust the belief space. There is no ϵ parameter in CURL because exploratory moves are provided by EC.

We may use a steady-state or generational GA, or other form of EC algorithm, and we may use one of the Q-Learning or Q(λ) algorithms to update the $Q(s, a)$, but in this paper we use the GENITOR-based NGA and SARSA(λ). The resulting algorithm is outlined in Figure 2, in which SARSA($\lambda, \alpha, 0$) denotes a SARSA(λ) episode with a given value of the α parameter, following the policy specified by chromosome O while updating the $Q(s, a)$ as usual. As in NGA the population in randomly initialised and fitness is evaluated using S samples. Note that for a deterministic POMDP only one sample is needed to obtain the fitness of a chromosome, so we can set $S = 1$ to obtain a CURL hybrid of SARSA(λ) and GENITOR.

```
CURL (S, P, pc, pm, α, λ, pl) :
(   create population of size P
    evaluate population using S samples
    initialise the Q(s,a)
    while not(termination condition)
    (   generate an offspring O using pc, pm
        update O using pl and the Q(s,a)
        call SARSA(λ,α,O) S times to estimate O fitness
            and bootstrap the Q(s,a)
        replace least-fit chromosome by O
    )
    output fittest chromosome
)
```

Fig. 2. CURL instantiation

2.2 Convergence

For POMDPS, unlike MDPs, suboptimal policies can form local optima in policy space [20]. This motivates the use of global search techniques such as EC, which are less likely to become trapped in local optima, and a hybrid such as CURL uses EC to directly explore policy space. CURL also uses bootstrapping to perform small changes to the policy by hill-climbing on the $Q(s, a)$ values. Hill-climbing has often been combined with GAs to form *memetic algorithms* with faster convergence than a pure GA, and this was a motivation for CURL's design. However, if bootstrapping is used then optimal policies are not necessarily stable: that is, an optimal policy might not attract the algorithm [20]. Thus a hybrid might not be able to find an optimal policy even if it escapes all local optima. The possible instability of optimal policies does not necessarily render such hybrids useless, because there might be optimal or near-optimal policies that *are* stable, but convergence is a very desirable property.

Fortunately, it is easy to show that if $p_l < 1$ and the underlying EC algorithm is convergent then so is CURL: if $p_l < 1$ then there is a non-zero probability that no allele is modified by the $Q(s, a)$, in which case CURL behaves exactly like the EC algorithm. This is not true of all hybrids (for example [9]). The GA used in the CURL instantiation is convergent (to within some accuracy depending on the number of samples used), because every gene in a new chromosome can potentially be mutated to an arbitrary allele. Therefore the CURL instantiation is convergent.

2.3 Note

Ideally, we should now evaluate CURL on standard POMDPs from the literature, but we shall postpone this for future work. The work in this paper is motivated by the need to solve large, complex inventory control problems that do not succumb to more traditional methods. In fact we know of no method in the inventory control literature that can optimally solve our problem in a reasonable time (at least, the constrained form of the problem: see below). We shall therefore test CURL on POMDPs from stochastic inventory control. We believe that the problem we tackle has not previously been considered as a POMDP, but we shall show that it is one.

3 POMDPs from Stochastic Inventory Control

The problem is as follows. We have a planning horizon of N periods and a demand for each period $t \in \{1, \dots, N\}$, which is a random variable with a given probability density function; we assume that these distributions are normal. Demands occur instantaneously at the beginning of each time period and are *non-stationary* (can vary from period to period), and demands in different periods are independent. A fixed delivery cost a is incurred for each order (even for an order quantity of zero), a linear holding cost h is incurred for each product unit carried in stock from one period to the next, and a linear stockout cost s is incurred for each period in which the net inventory is negative (it is not possible to sell back excess items to the vendor at the end of a period). The aim is to find a replenishment plan that minimizes the expected total cost over the planning

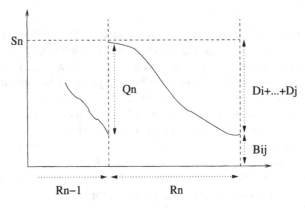

Fig. 3. The (R, S) policy

horizon. Different inventory control policies can be adopted to cope with this and other problems. A policy states the rules used to decide when orders are to be placed and how to compute the replenishment lot-size for each order.

3.1   Replenishment Cycle Policy

One possibility is the *replenishment cycle policy* (R, S). Under the non-stationary demand assumption this policy takes the form (R^n, S^n) where R^n denotes the length of the nth replenishment cycle and S^n the order-up-to-level for replenishment. In this policy a strategy is adopted under which the actual order quantity for replenishment cycle n is determined only after the demand in former periods has been realized. The order quantity is computed as the amount of stock required to raise the closing inventory level of replenishment cycle $n - 1$ up to level S^n. To provide a solution we must populate both the sets R^n and S^n for $n = \{1, \ldots, N\}$. The (R, S) policy yields plans of higher cost than the optimum, but it reduces planning instability [7] and is particularly appealing when items are ordered from the same supplier or require resource sharing [27]. Figure 3 illustrates the (R, S) policy. R^n denotes the set of periods covered by the nth replenishment cycle; S^n is the order-up-to-level for this cycle; Q_n is the expected order quantity; $D_i + \ldots + D_j$ is the expected demand; B_{ij} is the buffer stock required to guarantee service level α.

Though both RL and EC have been applied to a variety of inventory control problems, some of them POMDPs [31], neither seems to have been applied to this important problem. There are more efficient algorithms which are guaranteed to yield optimal policies (under reasonable simplifying assumptions) so RL and EC would not be applied to precisely this problem in practice. However, if we complicate the problem in simple but realistic ways, for example by adding order capacity constraints or dropping the assumption of independent demands, then these efficient algorithms become unusable. In contrast, RL and EC algorithms can be used almost without modification. Thus the problem is useful as a representative of a family of more complex problems.

Note that the inventory control term *policy* refers to the *form* of plan that we search for (such as the (R, S) policy), whereas a POMDP policy is a *concrete* plan (such as the (R, S) policy with given (R^n, S^n) values). We use the term in both senses but the meaning should be clear from the context.

3.2 POMDP Model

The replenishment cycle policy can be modelled as a POMDP as follows. Define a *state* to be the period n, an *action* to be either the choice of an order-up-to level or the lack of an order (denoted here by a special action N), and a *reward* r_n to be minus the total cost incurred in period n. The rewards are *undiscounted* (do not decay with time), the problem is *episodic* (has well-defined start and end states), the POMDP is *nondeterministic* (the rewards are randomised), and its solution is a policy that is *deterministic* and *memoryless* (actions are taken solely on the basis of the agent's current observations). This problem is non-Markovian but has an underlying MDP. Suppose we include the current stock level (suitably discretised or approximated) in the state. We then have the Markov property: the current stock level and period is all the information we need to make an optimal decision. But the (R, S) policy does not make optimal decisions: instead it fixes order-up-to levels independently of the stock level.

The problem is slightly unusual as a POMDP for two reasons. Firstly, all actions from a state n lead to the same state $n + 1$ (though they have different expected rewards): different actions usually lead to different states. Secondly, many applications are non-Markovian because of limited available information, but here we *choose* to make it non-Markovian by discarding information for an application-specific reason: to reduce planning instability. Neither feature invalidates the POMDP view of the problem, and we believe that instances of the problem make ideal benchmarks for RL and EC methods: they are easy to describe and implement, hard to solve optimally, have practical importance, and it turns out that neither type of algorithm dominates the other.

There exist techniques for improving the performance of RL algorithms on POMDPs, in particular the use of forms of memory such as a belief state or a recurrent neural network. But such methods are inapplicable to our problem because the policy would not then be memoryless, and would therefore not yield a replenishment cycle policy. The same argument applies to stochastic policies, which can be arbitrarily more efficient than deterministic policies [28]: for our inventory problem we require a deterministic policy. Thus some powerful RL techniques are inapplicable to our problem.

4 Experiments

We compare SARSA(λ), the NGA and CURL on five benchmark problems. The instances are shown in Table 1 together with their optimal policies. Each policy is specified by its planning horizon length R and its order-up-to-level S, and the expected cost of the policy per period is also shown, which can be multiplied by the number of periods to obtain the expected total cost of the policy. For example instance (3) has the optimal policy $[159, N, N, 159, N, N, 159, \ldots]$. However, the policy is only optimal if the total number of periods is a multiple of R, and we choose 120 periods as a common multiple of $R \in \{1, 2, 3, 4, 5\}$. This number is also chosen for hardness: none of the three

Table 1. Instances and their optimum policies

#	h	s	a	demand mean	demand std dev	R		S	cost/ period	cost/120 periods
(1)	1	10	50	50	10	1	63	68	8160
(2)	1	10	100	50	10	2	112	94	11280
(3)	1	10	200	50	10	3	159	138	16560
(4)	1	10	400	50	10	4	200	196	23520
(5)	1	10	800	50	10	5	253	279	33480

algorithms find optimal policies within 10^8 simulations (a Mixed Integer Programming approach also failed given several hours). We varied only the a parameter, which was sufficient to obtain different R values (and different results: see below). We allow 29 different order-up-to levels at each period, linearly spaced in the range 0–280 at intervals of 10, plus the N no-order option, so from each state we must choose between 30 possible actions. This range of order-up-to levels includes the levels in the optimum policies for all five instances. Of course if none of the levels coincides with some order-up-to-level in an optimal policy then this prevents us from finding the exact optimum policy. But even choosing levels carefully so that the exact values are reachable does not lead to optimal policies using the three algorithms.

As mentioned above, this problem can be solved in polynomial time because of its special form, which is how we know the optimum policies. We therefore also generate five additional instances (1c,2c,3c,4c,5c) by adding an order capacity constraint to instances (1,2,3,4,5) respectively, simply by choosing an upper bound below the level necessary for the optimum policy. For each instance the 30 levels are linearly spaced between 0 and $\lfloor 0.8S \rfloor$ (respectively 54, 89, 127, 156 and 223). This problem is NP-hard and we know of no method that can solve it to optimality in a reasonable time. We therefore do not know the optimum policies for these instances, only that their costs are at least as high as those without the order constraints.

We tailored NGA and CURL for our application by modifying the mutation operator: because of the special nature of the N action we mutate a gene to N with 50% probability, otherwise to a random order-up-to level. This biased mutation improves NGA and CURL performance. We also tailored SARSA and CURL for our application. Firstly, we initialise all state-action values to the optimistic value of 0, because the use of optimistic initial values encourages early exploration [30]. Secondly, we experimented with different methods for varying ϵ, which may decay with time using different methods. [3,16] decrease ϵ linearly from 0.2 to 0.0 until some point in time, then fix it at 0.0 for the remainder. [30] recommend varying ϵ inversely with time or the number of episodes. We found significantly better results using the latter method, under the following scheme: $\epsilon = 1/(1 + \epsilon' e)$ where e is the number of episodes so far and ϵ' is a fixed coefficient chosen by the user. For the final 1% of the run we set $\epsilon = 0$ so that the final policy cost reflects that of the greedy policy (after setting ϵ to 0 we found little change in the policy, so we did not devote more time to this purely greedy policy).

Each of the three algorithms has several parameters to be tuned by the user. To simulate a realistic scenario in which we must tune an algorithm once then use it many times,

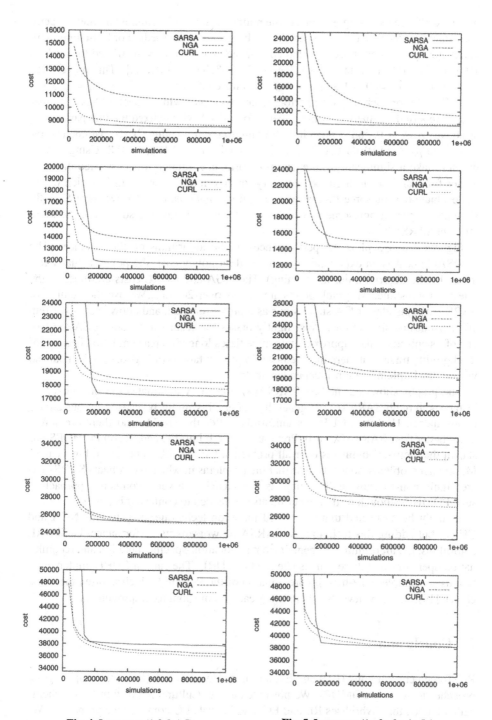

Fig. 4. Instances (1,2,3,4,5)   **Fig. 5.** Instances (1c,2c,3c,4c,5c)

we tuned all three to a single instance: the middle instance (3) without an order capacity constraint. For SARSA(λ) we tuned $\epsilon', \alpha, \lambda$ by the common method of hill-climbing in parameter space to optimise the final cost of the evolved policy, restricted to λ values $\{0.0, 0.1, \ldots, 0.9, 1.0\}$ and ϵ', α values $\{0.1, 0.03, 0.01, 0.003, \ldots\}$. This process led to $\alpha = 0.003$, $\epsilon' = 0.001$ and $\lambda = 0.7$. We chose NGA settings $p_c = p_m = 0.5$ and $P = S = 30$ for each instance: performance was robust with respect to these parameters, as reported by many GA researchers. To tune CURL we fixed the GA parameters as above, set $\lambda = 0$, and applied hill-climbing to the remaining CURL parameters, restricted to $p_l \in \{1.0, 0.3, 0.1, 0.03, \ldots\}$, to obtain $\alpha = 0.1$, $p_l = 0.3$. Using $\lambda > 1$ did not make a significant difference to performance (though it necessitated different values for α and p_l): it might be necessary for deterministic problems in which we do not evaluate chromosome fitness over several simulations, but here we have $S = 30$ simulations per chromosome in which to perform bootstrapping so we use the more efficient SARSA(0).

Figures 4 and 5 plots the performances of the algorithms on the instances. The SARSA(λ) cost is an exponentially-smoothed on-policy cost (the policy actually followed by the algorithm during learning). The NGA and CURL costs are those of the fittest chromosome. All graph points are means over 20 runs. We use the number of SARSA(λ) episodes or GA simulations as a proxy for time, and allow each algorithm 10^6 episodes or simulations. This slightly biases the results in favour of SARSA(λ): one of its episodes takes approximately three times longer than a simulation because of its eligibility trace. But there may be faster implementations of SARSA(λ) than ours so we use this implementation-independent metric.

The graphs show that neither SARSA(λ) nor NGA dominates the other over all instances, though SARSA(λ) is generally better (this might be caused by our choice of instances). However, CURL is uniformly better than NGA, and therefore sometimes better than SARSA(λ) also. Previous research into EC and RL on POMDPS has shown that neither dominates over all problems, but that EC is better on highly non-Markovian problems, so we assume that the problems in which NGA beats SARSA(λ) are highly non-Markovian. This implies that CURL is a very promising approach to such POMDPs, though further experiments are needed to confirm this pattern.

It might be suspected that the biased mutation technique unfairly aids NGA and CURL: but adding this technique to SARSA(λ) worsens its performance. Unlike RL algorithms, EC algorithms can benefit from application-specific mutation and recombination operators, and these can also be used in CURL. The current CURL implementation uses a simple table-based representation of the POMDP, which is often the worst choice [19], so we believe that there is a great deal of room for improvement.

5   Conclusion

Reinforcement Learning (RL) and Evolutionary Computation (EC) are competing approaches to solving POMDPs. We presented a new Cultural Algorithm (CA) schema called CURL that hybridises RL and EC, and inherits EC convergence properties. We also described POMDPs from stochastic inventory theory on which neither RL nor EC dominates the other. In experiments a CURL instantiation outperforms the EC algorithm, and on highly non-Markovian instances it also outperforms the RL algorithm. We believe that CURL is a promising approach to solving POMDPs, combining EC and RL algorithms with little modification.

This work is part of a series of studies in solving inventory problems using systematic and randomised methods. In future work we intend to develop CURL for more complex inventory problems, and for more standard POMDPs from the Artificial Intelligence literature.

Acknowledgement. This material is based in part upon works supported by the Science Foundation Ireland under Grant No. 05/IN/I886, and under Grant No. 03/CE3/I405 as part of the Centre for Telecommunications Value-Chain-Driven Research (CTVR). S. A. Tarim and B. Hnich are supported by the Scientific and Technological Research Council of Turkey (TUBITAK) under Grant No. SOBAG-108K027.

References

1. Arnold, D.V., Beyer, H.-G.: Local Performance of the (1+1)-ES in a Noisy Environment. IEEE Trans. Evolutionary Computation 6(1), 30–41 (2002)
2. Becerra, R.L., Coello, C.A.C.: A Cultural Algorithm with Differential Evolution to Solve Constrained Optimization Problems. In: Lemaître, C., Reyes, C.A., González, J.A. (eds.) IBERAMIA 2004. LNCS (LNAI), vol. 3315, pp. 881–890. Springer, Heidelberg (2004)
3. de Croon, G., van Dartel, M.F., Postma, E.O.: Evolutionary Learning Outperforms Reinforcement Learning on Non-Markovian Tasks. In: Workshop on Memory and Learning Mechanisms in Autonomous Robots, 8th European Conference on Artificial Life, Canterbury, Kent, UK (2005)
4. Fitzpatrick, J.M., Grefenstette, J.J.: Genetic Algorithms in Noisy Environments. Machine Learning 3, 101–120 (1988)
5. Gao, F., Cui, G., Liu, H.: Integration of Genetic Algorithm and Cultural Algorithms for Constrained Optimization. In: King, I., Wang, J., Chan, L.-W., Wang, D. (eds.) ICONIP 2006. LNCS, vol. 4234, pp. 817–825. Springer, Heidelberg (2006)
6. Gopalakrishnan, G., Minsker, B.S., Goldberg, D.: Optimal Sampling in a Noisy Genetic Algorithm for Risk-Based Remediation Design. In: World Water and Environmental Resources Congress. ASCE (2001)
7. Heisig, G.: Comparison of (s,S) and (s,nQ) Inventory Control Rules with Respect to Planning Stability. International Journal of Production Economics 73, 59–82 (2001)
8. Holland, J.H.: Adaptation. In: Progress in Theoretical Biology IV, pp. 263–293. Academic Press, London (1976)
9. Iglesias, R., Rodriguez, M., Sánchez, M., Pereira, E., Regueiro, C.V.: Improving Reinforcement Learning Through a Better Exploration Strategy and an Adjustable Representation of the Environment. In: 3rd European Conference on Mobile Robots (2007)
10. Jaakkola, T., Singh, S.P., Jordan, M.I.: Reinforcement Learning Algorithm for Partially Observable Markov Decision Problems. In: Advances in Neural Information Processing Systems 6. MIT Press, Cambridge (1994)
11. Kovacs, T., Reynolds, S.I.: A Proposal for Population-Based Reinforcement Learning. Technical report CSTR-03-001, Department of Computer Science, University of Bristol (2003)
12. Littman, M.: Memoryless Policies: Theoretical Limitations and Practical Results. In: 3rd Conference on Simulation of Adaptive Behavior (1994)

13. Littman, M.L., Cassandra, A.R., Kaelbling, L.P.: Learning Policies for Partially Observable Environments: Scaling Up. In: International Conference on Machine Learning (1995)
14. Littman, M., Dean, T., Kaelbling, L.: On the Complexity of Solving Markov Decision Problems. In: 11th Conference on Uncertainty in Artificial Intelligence, pp. 394–402 (1995)
15. Liu, H., Hong, B., Shi, D., Ng, G.S.: On Partially Observable Markov Decision Processes Using Genetic Algorithm Based Q-Learning. In: Advances in Neural Networks, pp. 248–252. Watam Press (2007)
16. Loch, J., Singh, S.P.: Using Eligibility Traces to Find the Best Memoryless Policy in Partially Observable Markov Decision Processes. In: 15th International Conference on Machine Learning, pp. 323–331 (1998)
17. Miller, B.L.: Noise, Sampling, and Efficient Genetic Algorithms. PhD thesis, University of Illinois, Urbana-Champaign (1997)
18. Miller, B.L., Goldberg, D.E.: Optimal Sampling for Genetic Algorithms. In: Intelligent Engineering Systems Through Artificial Neural Networks, vol. 6, pp. 291–298. ASME Press (1996)
19. Moriarty, D.E., Schultz, A.C., Grefenstette, J.J.: Evolutionary Algorithms for Reinforcement Learning. Journal of Artificial Intelligence Research 11, 241–276 (1999)
20. Penrith, M.D., McGarity, M.J.: An Analysis of Direct Reinforcement Learning in non-Markovian Domains. In: 15th International Conference on Machine Learning, pp. 421–429. Morgan Kaufmann, San Francisco (1998)
21. Reynolds, R.G.: An Introduction to Cultural Algorithms. In: 3rd Annual Conference on Evolutionary Programming, pp. 131–139. World Scientific Publishing, Singapore (1994)
22. Reynolds, R.G.: Cultural Algorithms: Theory and Applications. New Ideas in Optimization, pp. 367–377. McGraw-Hill, New York (1999)
23. Reynolds, R.G., Chung, C.: A Cultural Algorithm Framework to Evolve Multiagent Cooperation With Evolutionary Programming. In: Angeline, P.J., McDonnell, J.R., Reynolds, R.G., Eberhart, R. (eds.) EP 1997. LNCS, vol. 1213, pp. 323–333. Springer, Heidelberg (2006)
24. Rivera, D.C., Becerra, R.L., Coello, C.A.C.: Cultural Algorithms, an Alternative Heuristic to Solve the Job Shop Scheduling Problem. Engineering Optimization 39(1), 69–85 (2007)
25. Rummery, G.A., Niranjan, M.: On-line Q-learning Using Connectionist Systems. Technical report CUED/F-INFENG/TR 166, Cambridge University (1994)
26. Russell, S.J., Zimdars, A.: Q-Decomposition for Reinforcement Learning Agents. In: 20th International Conference on Machine Learning, pp. 656–663. AAAI Press, Menlo Park (2003)
27. Silver, E.A., Pyke, D.F., Peterson, R.: Inventory Management and Production Planning and Scheduling. John-Wiley and Sons, New York (1998)
28. Singh, S., Jaakkola, T., Jordan, M.: Learning Without State-Estimation in Partially Observable Markovian Decision Processes. In: 11th International Conference on Machine Learning, pp. 284–292. Morgan Kaufmann, San Francisco (1994)
29. Stroud, P.D.: Kalman-Extended Genetic Algorithm for Search in Nonstationary Environments with Noisy Fitness Functions. IEEE Transactions on Evolutionary Computation 5(1), 66–77 (2001)
30. Sutton, R.S., Barto, A.G.: Reinforcement Learning: An Introduction. MIT Press, Cambridge (1998)
31. Treharne, J.T., Sox, C.R.: Adaptive Inventory Control for Nonstationary Demand and Partial Information. Management Science 48(5), 607–624 (2002)
32. Watkins, C.J.C.H.: Learning From Delayed Rewards. PhD thesis, Cambridge University (1989)
33. Whitley, D., Kauth, J.: GENITOR: A Different Genetic Algorithm. In: Rocky Mountain Conference on Artificial Intelligence, Denver, CO, USA, pp. 118–130 (1988)

A Variable Neighborhood Search Integrated in the POPMUSIC Framework for Solving Large Scale Vehicle Routing Problems

Alexander Ostertag, Karl F. Doerner, and Richard F. Hartl

Department of Business Administration, University of Vienna,
Bruenner Strasse 72, 1210 Vienna, Austria
{Alexander.Ostertag,Karl.Doerner,Richard.Hartl}@univie.ac.at

Abstract. This paper presents a heuristic approach based on the POP-MUSIC framework for solving large scale Multi Depot Vehicle Routing Problems with Time Windows derived from real world data. A Variable Neighborhood Search is used as the optimizer in the POPMUSIC framework. POPMUSIC is a new decomposition approach for large scale problems. We compare our method with a pure VNS approach and a Memetic Algorithm integrated in a POPMUSIC framework. The computational results show that the integration of VNS in the POPMUSIC framework outperforms the other existing methods. Furthermore different distance metrics for the decomposition strategies are presented and the results are reported and analyzed.

Keywords: Vehicle Routing, Variable Neighborhood Search, Problem Decomposition.

1 Introduction

In this paper we present a solution method based on the POPMUSIC framework to efficiently solve real world problems of large scale. We used the Variable Neigborhood Search (VNS) as the heuristic optimizer in the framework. The contribution of this paper is twofold. First we propose an efficient integration of the VNS in the POPMUSIC framework. The designed solution method is highly efficient and outperforms the existing methods for solving large scale Multi Depot Vehicle Routing Problems with Time Windows (MDVRPTW) published in Ostertag et al. [13]. Second we design and analyze different metrics for the generation of sub-problems. Decomposing a large scale real world routing problem - especially the MDVRPTW - in several sub-problems is not trivial. A detailed numerical analysis is provided to evaluate the promising decomposition strategies.

The MDVRPTW is a generalization of the Vehicle Routing Problem (VRP) which is known to be NP-hard. Multiple depots as well as time windows are added to the classical VRP in order to receive a more realistic model for the problems of modern carrier fleet companies. As the problem is NP-hard only small instances can be solved to optimality in reasonable time. Therefore metaheuristic

M.J. Blesa et al. (Eds.): HM 2008, LNCS 5296, pp. 29–42, 2008.

approaches are developed to provide good quality solutions at reasonable computational effort. Some of these metaheuristics scale incredibly well with problem
sizes of up to a couple of hundred customers, however only a few exist that can
be applied to large scale real world problems. Homberger and Gehring [8] show
that large instances of up to 1000 customers can be solved with their 2-phase
hybrid metaheuristic approach. An active guided evolution strategy by Mester
and Bräysy [12, 11] as well as a VNS approach by Kytöjoki et al. [9] were also
developed to solve instances of larger sizes. Alternative strategies like the POP-
MUSIC framework by Taillard and Voss [18] focus on intelligently decomposing
the problem into smaller sub-problems that are then solved sequentially by well
known methods. The decomposition phase is repeated iteratively. Flaberg et al.
[3] also successfully applied a decomposition strategy to solve a large scale news-
paper delivery problem. The MDVRPTW is up to now not so well studied. The
most recent developed methods for the problem at hand are the tabu search by
Cordeau et al. [2] and the VNS by Polacek et al. [15], however these methods
are only applied on instances of moderate size.

The remainder of the paper is organized as follows. The problem is explained in
detail in Section 2. Section 3 gives an overview of the VNS used as an optimizer.
The POPMUSIC framework is described in Section 4. In Section 5 the numerical
results are reported. In this section also different decomposition strategies are
analyzed. In the conclusion (Section 6) we summarize the results and provide
ideas for further research.

2   Problem Description

The real world problem considered in this paper was slightly modified so that
it can be considered as a MDVRPTW. Customers that could no be reached in
time, as well as a handful of backhauls were purged from the data set. Compared
to the well known Vehicle Routing Problem with Time Windows (VRPTW), the
MDVRPTW is extended by having more than one depot with different locations
and associated vehicle fleets. The MDVRPTW is defined on a complete graph
$G = (V, A)$ where $V = \{v_1, ..., v_m, v_{m+1}, ..., v_{m+n}\}$ is the vertex set and $A =$
$\{(v_i, v_j) : v_i, v_j \in V, i \neq j\}$ is the arc set. The n customers are represented by
vertices v_{m+1} to v_{m+n}, while v_1 to v_m stand for the m depots. Several weights
are associated to each vertex $v_i \in V, i = m+1, ..., m+n$. These weights represent
the demands d_i, the service times s_i, as well as the time windows $[e_i, l_i]$ which
are defined by the earliest e_i and latest l_i possible start times for the service.
These time windows also apply to the depots $(i = 1, ..., m)$ where they define
the opening hours of the depots. Each arc (v_i, v_j) is associated with a non-
negative travel time or cost. A vehicle fleet consisting of a total of K vehicles is
globally assigned to the m depots. The fleet is homogeneous and each vehicle is
characterized by a non-negative capacity D and a non-negative maximum route
duration T. Finally, the distribution of the vehicles over the depots is defined as
input data. The aim is to build K vehicle routes, each route starting from a depot
and returning back to the same depot, so that each customer i belongs to exactly

one route and is serviced during its corresponding time window $[e_i, l_i]$. Each vehicle route has to satisfy the additional constraints of the maximal allowed tour length T and vehicle capacity D. The objective considered in this paper is to minimize the total distance traveled by all vehicles.

3 Variable Neighborhood Search

The VNS [6] used was derived from the work done by Polacek et al. [15] as it has been successfully applied to MDVRPTW problem instances. Even though other methods like TS, iterative local search, ACO or hill climbing may be interesting to implement into the POPMUSIC framework we did choose the VNS because the most recent VNS approaches outperformed other techniques for the Vehicle Routing Problems see [4], [7] or [14] and especially [15]. The VNS consists of four phases (construct initial solution, shaking, iterative improvement, acceptance decision). Phases two to four are iteratively executed.

3.1 Construct Initial Solution

We use a modified Clarke and Wright Savings algorithm [1] to generate the initial solution. Customers are assigned to empty routes with starting point and endpoint at the closest depot. We then generate a sorted list according to the savings value that can be realized when merging two partial routes. In order to generate different starting solutions for the different runs a stochastic component is introduced; with a probability of 0.9 the highest possible savings value is realized. The algorithm terminates when no more routes can be feasible merged. The constructed solution is the first incumbent solution within the VNS.

3.2 Shaking

A very important design decision for the VNS is the selection of the right neighborhood structure in the shaking phase. A CROSS-Operator [17] as well as an inverted CROSS operator are used to perturb the incumbent solution. The multi depot feature of the real world problem is considered by defining on which routes the CROSS-Operator is applied. Two different variants are realized. In the first variant only routes belonging to the same depot may be changed. In the second variant routes starting at different depots may be changed. In both of the mentioned variants the maximal allowed sequence length that may be changed by the CROSS operator is defined between zero and five. An additional case is considered where the maximum allowed length is equal to the number of customers in the route with the smaller number of customers of the two considered routes. The two routes, on which the operator is applied, are selected at random.

3.3 Iterative Improvement

After the shaking phase each solution is improved by an iterative improvement procedure. In this procedure a restricted 3-opt [10] is used as local search operator. The operator is restricted to a maximum allowed sequence length of three customers. We realized a first improvement strategy, which means the algorithm accepts the current solution as new incumbent solution as soon as an improvement is found and restarts the iterative improvement procedure. A special feature of real world problems is that customers are often on the same geographic location (hospital, shopping mal, business centers,..) . For this fact a restrictive sequence length is not useful. It can happen that more than five customers have to be visited on the same day within the same shopping mal. When these customers are on the same route the 3-opt operator has no effect. We therefore allow within the 3-opt procedure to shift customers on the same location without any restriction.

3.4   Acceptance Decision

The fitness evaluation function of a solution S follows the implementation of [2] and [15]. The total travel time of the routes is denoted by $c(S)$. The values $q(S)$, $t(S)$ and $w(S)$ respectively denote the total violation of load, duration and time window constraints. The arrival time a_i at each customer i is calculated and an arrival after the end of the time window $a_i > l_i$ is penalized while an arrival before the start of the time window $a_i < e_i$ is allowed but generates a waiting time. Each route is then checked for violations with respect to D and T as well as the total violation of the time window constraints $\sum_{i=1}^{n} max(0, a_i - l_i)$. The fitness function is defined by $f(S) = c(S) + \alpha q(S) + \beta t(S) + \gamma w(S)$ where α, β and γ are positive weights which are all set to 100 to strongly penalize infeasibility.

To overcome the problem of getting stuck in local optima, we sometimes allow inferior solutions to be accepted. While better solutions are always accepted, worse solutions are only allowed when two criteria are met. The first criterion defines that we generally allow deteriorating solutions after a certain number of unproductive iterations (10^5). If the limit of unproductive iterations is reached the next generated solution with a deteriorating fitness value that fulfills the second criterion is accepted. The second criterion defines the threshold value for accepting a deteriorating solution. The threshold is defined by a ratio (in our experiments 5%) of the best found solution so far to the current solution. The counter that measures the unproductive iterations is set to zero as soon as a deteriorating solution is accepted.

4   POPMUSIC

Taillard and Voss [18] propose the POPMUSIC framework for dealing with problems of large size. The basic concept of the framework is to decompose a pre-calculated solution S into p parts $s_1, ..., s_p$. In the next step some of these parts are aggregated into a sub-problem. An optimizer then tries to improve this sub-problem. If parts and sub problems are well defined, every improvement of a sub problem corresponds to an improvement of the whole solution S. The described process is then repeated and different sub-problems are build and optimized. The basic POPMUSIC framework is described in algorithm 1.

Algorithm 1. Basic POPMUSIC framework

Input: Solution S composed of parts $s_1, ..., s_p$, parameter r
Set $A \leftarrow \emptyset$
while $A \neq \{s_1, ..., s_p\}$ **do**
   Select seed part $s_i \notin A$
   Create a sub-problem R_i composed of the r parts $s_{i_1}, ..., s_{i_r}$ most related to s_i
   Optimize R_i
   if R_i has been improved **then**
      Update S (and corresponding parts)
      Set $A = A \setminus \{s_{i_1}, ..., s_{i_r}\}$
   else
      Set $A \leftarrow A \cup \{s_i\}$
   end if
end while

4.1 POPMUSIC Customization

To apply POPMUSIC to the real world MDVRPTW problem at hand, it is necessary to specify its principal components. We defined a part $(s_1, ..., s_p)$ as a specific route in the complete solution. Therefore the applied proximity measure (*relation - most related to s_i*) needs to measure the distance between two routes so that appropriate sub problems can be created. We examined two different ways to measure proximity that will be explained in detail in the following subsections.

We therefore define a sub problem as a subset of r routes that can be treated and solved like an independent MDVRPTW of desired size. We decided to set r to 10 routes as the VNS provides good solutions in reasonable computation time for problems of moderate size. This was also shown in [2, 15]. All resulting sub problems are then improved by applying the VNS. The VNS terminates either if an iteration limit is reached or the computational time exceeds a certain threshold. Since the resulting sub-problems depend strongly on the selection of the seed customer or part, seed parts are chosen in a systematic way.

4.2 Proximity Measures

We decided to test two different relatedness measures for potential parts of a new sub-problem to see which will result in a better decomposition of the problem. The two measures are different in the way the distance is measured, we measure distance by travel time and through the use of trigonometric functions (the sweep idea [5]). Furthermore we tried a total of eight different strategies to measure proximity, that will be explained in detail in the following subsection. The three measures based on the sweep idea are presented before the five measures based on the distance.

Proximity Through Sweeping. This measure defines distance by the angle between the centers of two routes. The centers of these two routes are defined as the centers of gravity of all the customers in the according route. This concept

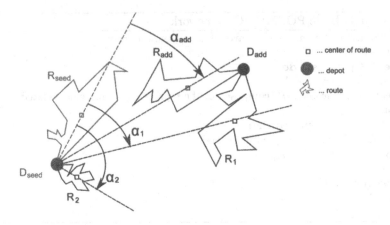

Fig. 1. Measure S_I (sweep with no restriction)

was also used in [16] and introduced in [17]. The angle α is than calculated between these two centers, with the pivot point being the depot (D_{seed}) of the seed route (R_{seed}).

Sweep with no restriction (S_I)

The most basic way to apply this measure is without restricting the selection of the routes that can be added to create the sub-problem. It is therefore created by adding the routes with the smallest angle up to the maximum allowed sub-problem size r. In order to allow some diversity in the creation of the sub problem we use a probability of 0.1 that a selected route is rejected and cannot enter the sub problem. The procedure is depicted in Figure 1.

Sweep with tight restriction (S_{II}).

The sweep selection of routes may be not suitable to deal with multiple depots as it can happen that a route gets selected even when a couple of other depots are in-between. We therefore restricted the selection of routes that belong to another depot than the seed part in the following way. Routes are still added with the smallest angle to the seed part, however they are only added when one of the two following criteria is fulfilled.

1. The distance between the route to be added and the seed depot is smaller than the distance between the two depots.
2. The route to be added is closer to the seed depot than to the original depot.

The procedure is depicted in Figure 2. Routes can only be selected in the grey area, which is shown divided for each criterion.

Sweep with loose restriction (S_{III}).

We relaxed the criteria of S_{II} in a way that routes to be added that belong to another depot may be selected when their center of gravity does not lie behind the second depot (D_{add}) by a certain angle β.

The procedure is depicted in Figure 3. Routes can only be selected in the grey area.

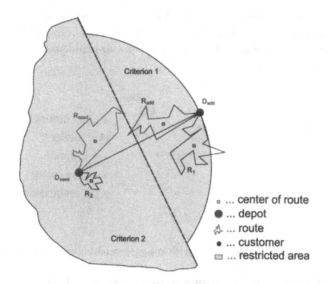

Fig. 2. Measure S_{II} (sweep with tight restriction)

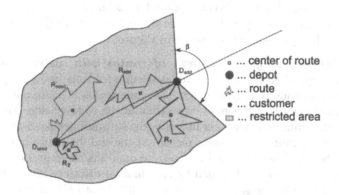

Fig. 3. Measure S_{III} (sweep with no restriction)

Proximity by Smallest Distance. In the second group of measures the euclidian distance between two entities is used as proximity. Entities can be single customers or all customers in a route. When a group of customers form an entity then the center of gravity of these customers is used to calculate the distance. The setups of the different strategies vary in the aggregation level as well as on how customers are selected to join the sub-problem.

Distance between aggregated customers of routes (D_I). In this strategy the centers of gravities for each route define the entities and they are used for distance calculations of the neighboring routes. All the distances between the seed route (R_{seed}) and all possible other routes are computed and stored in a sorted list. Then 75 % of the routes with the shortest distance to the seed route (R_{seed}) are combined. The missing routes are then added through the use of a roulette wheel selection. Routes which are closer to the seed route have a higher probability of being selected. Routes are selected as long as the sub-problem contains r routes.

The strategy is illustrated in Figure 4.

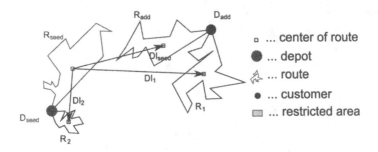

Fig. 4. Measure D_I (distance between aggregated customers of routes)

Distance between single customers of routes (D_{II}). Here we generate a list containing the distances between all customers of a seed route and all other remaining customers. The customers with the smallest distance are then added to the sub-problem, however because a part is defined as a route we add the complete route to the sub-problem. In total $r - 1$ routes are then added to the sub problem. The procedure is depicted in Figure 5.

Distance between single customers of routes with restriction (D_{III}). With the concept of strategy D_{II} generated sub-problems in highly populated regions like cities contain only routes in the same region. This happens because distances are smaller on average in cities than in the country side. Therefore as soon as a route contains a customer in a city, the sub-problem is always extended with customers or routes in the same region. At the end the routes in the country side cannot be combined reasonably. To overcome this structural drawback we modified D_{II}. Routes are still added by the smallest distance, however only one route per customer in the seed route may be added until all other customers in the seed route have added the same amount of routes.

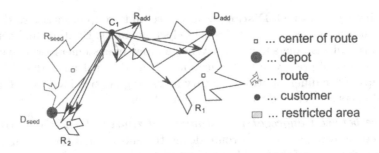

Fig. 5. Measure D_{II} (distance between single customers of routes)

Distance between aggregated customers of the seed route and a single customer (D_IV). Here we aggregated the customers in the seed part by calculating the centers of gravity of the route. Distances are than calculated between these center and each remaining customer of the other routes. The sub-problems are then build in standard manner. The procedure is depicted in Figure 6.

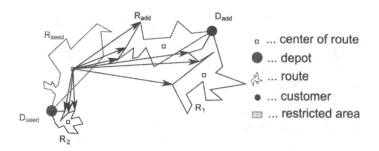

Fig. 6. Measure D_{IV} (distance between aggregated customers of the seed route and a single customer)

Distance between a single customer in the seed route and aggregated customers of a random route (D_V). Distances are calculated between a single customer in the seed route and the aggregated center of gravity of a random route. It is therefore the opposite of D_{IV}. The procedure is depicted in Figure 7.

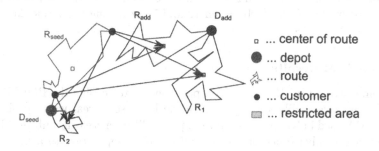

Fig. 7. Measure D_V (distance between a single customer in the seed route and aggregated customers of a random route)

5   Computational Experiments

5.1   Problem Characteristics

The problem considered originates from a large real world problem of an Austrian logistics provider that operates two distribution centers (depots) in or near the city of Vienna. The company serves from 700 to 2000 customers every day with

Table 1. Problems size

Day	1	2	3	4	5	6	7	8	9	10
Size	1201	1180	1284	1305	1175	743	889	1095	1848	1709

Table 2. Comparison of algorithms

Day	PopMA	VNS	PopVNS	RPD PopMA / PopVns	RPD VNS / PopVns
10	3990.30	3900.22	3617.17	-9.35%	-7.26%
11	4248.84	4337.41	4035.87	-5.01%	-6.95%
12	4337.51	4308.20	4002.35	-7.73%	-7.10%
13	4526.04	4641.33	4258.72	-5.91%	-8.24%
14	4335.46	4383.96	4085.11	-5.77%	-6.82%
22	2531.97	2592.75	2441.07	-3.59%	-5.85%
23	3483.54	3627.22	3394.80	-2.55%	-6.41%
24	3483.54	3354.47	3128.26	-10.20%	-6.74%
25	6031.33	5855.59	5368.09	-11.00%	-8.33%
26	5948.30	6344.28	5911.63	-0.62%	-6.82%
avg.	4291.68	4334.54	4024.31	-6.17%	-7.05%

a total number of 160 vehicles. Both depots are of equal size and the vehicle fleet is equally split between them. We evaluated 2 weeks, each with 5 days and customers between 743 and 1848 per day. Table 1 shows the number of customers to be served for each day. The customers have large time windows, some customers can be served in the morning between 8 and 12, some customers can be served in the afternoon between 12 and 16 or during the whole day from 8 to 16.

5.2   Numerical Results

It was demonstrated in [13] that decomposition strategies like the POPMUSIC framework are reasonable strategies to tackle large scale real world problems. While the MA used in [13] provides limited possibilities of creating sub-problems, especially in the inheritance of the individuals of a population to a newly decomposed problem, the VNS allows for more flexibility to create them. The MA uses a tournament selection method, a route based two-point crossover operator, a stochastic local search procedure for mutation and a steady state fashion updating of the population. It was used as the optimizer for the sub-problems that are generated by clusters of routes, so that information stored in the population can be taken from one subproblem to the next.

All the developed strategies were tested on Intel Pentium 640 'Prescott' 3,2 GHZ, 800 MHZ FSB, 2MB L2-Cache PCs. To study also the performance of our solution methods for more than two depots, we extended the initial problem to the three and four depot case. Ten independent runs with a computation time of five hours for each run were performed for each depot setup and strategy

Table 3. Results for two depots

measure	RPD to VNS	Rank
S_I	-6.62%	5
S_{II}	-6.62%	6
S_{III}	-6.56%	7
D_I	-6.87%	2
D_{II}	-6.71%	4
D_{III}	-6.85%	3
D_{IV}	-7.05%	1
D_V	-6.09%	8
MA	-0.80%	9
VNS	0.00%	10

Table 4. Results for three depots

measure	RPD to VNS	Rank
S_I	-6.38%	4
S_{II}	-5.97%	7
S_{III}	-6.45%	2
D_I	-6.14%	6
D_{II}	-6.17%	5
D_{III}	-6.41%	3
D_{IV}	-6.91%	1
D_V	-5.70%	8
VNS	0.00%	9

Table 5. Results for four depots

measure	RPD to VNS	Rank
S_I	-5.86%	3
S_{II}	-5.17%	7
S_{III}	-5.92%	2
D_I	-5.54%	5
D_{II}	-5.42%	6
D_{III}	-5.78%	4
D_{IV}	-6.31%	1
D_V	-5.13%	8
VNS	0.00%	9

combination. The VNS as an optimizer turns out to be able to outperform the MA [13] as an optimizer significantly. To be able to compare the two algorithms the code of the MA was used to run on the same instances and on the same machines. To present a fair comparison, the improved 3-opt used in the VNS was integrated in the MA. The improved 3-opt for the real-world case is able to shift customers on the same location regardless of the maximum allowed sequence length. Table 2 shows the average percentage deviation for the MA (integrated in the POPMUSIC framework) and the VNS without decomposition compared to the POPMUSIC VNS that uses the best performing proximity measure D_{IV}. We can only show the results for the two depot case, as the MA was not developed for more than two depots. The VNS integrated in POPMUSIC improves the average results compared to the MA integrated in POPMUSIC by 6.17 %.

5.3   Analysis of Proximity Measure

It can be seen that all of the tested measures perform significantly better than the VNS without decomposition (see tables 3,4,5). Over all instances and depot setups, D_{IV} provided the best results, except for the small instances in the two depot and four depot setup. It is closely followed by D_{III} which generated the second best results as can be seen in the overview in table 6. Concerning the sweep proximity measures S_{III} seems to work the best out of all tested sweep measures. Furthermore it is interesting to point out that S_{III} seems to gain efficiency when dealing with more depots. For the three and four depot case S_{III} provides the second best results. However D_{IV} provides also for the multi-depot case (three and four depots) the best results.

The results for the two depot case are presented in table 3, table 4 shows the results for the three depot case while table 5 shows the results for the four depot case. It can be seen that improvements by 6.56 % to 7.05 % can be realized for the two depot case when using POPMUSIC (see table 3). For the four depot case improvements between 5.13 % to 6.31 % are possible (see table 5). The average results over all depot cases are represented in table 6. It can be seen that the distance based strategies D_{IV} and D_{III} provides best results averaged on all the different problem settings. However also the sweep based strategy S_{III} provides almost similar results on average as D_{III} (see table 6).

Table 6. Results averaged over all depots

measure	RPD to VNS	Rank
S_I	-6.29%	4
S_{II}	-5.92%	7
S_{III}	-6.31%	3
D_I	-6.18%	5
D_{II}	-6.10%	6
D_{III}	-6.35%	2
D_{IV}	-6.76%	1
D_V	-5.64%	8
VNS	0.00%	9

6   Conclusion

In this paper we have shown that the POPMUSIC decomposition strategy that uses a VNS as sub problem optimizer can solve large scale MDVRPTW. It is shown that the results provided by the existing MA can be improved by roughly 6.17 % in the two depot setup. The results can be improved by 6.76 % over all instances when using POPMUSIC and VNS in comparison to a pure VNS approach.

We presented a number of different ways to measure proximity in an environment with a large amount of customers and more than one depot in order to decompose a large routing problem. The results show that the distance based proximity measures provide the best results (especially strategy D_{III} and D_{IV}), while properly implemented sweep based measures only work well when dealing with a higher number of depots (especially strategy S_{III}). We want to point out that the worst decomposing strategy presented performs still 5 % better than using no decomposition at all. This leads to the assumption that decomposition improves the solution quality when tackling large scale problems with current state-of-the art methods and computers on the basis of the same runtime. The POPMUSIC framework is easy to develop for large VRP instances and sub problem optimizers based on metaheuristic concepts and can easily be integrated to further improve solution quality. Especially local search based concepts that work with one incumbent solution, e.g. VNS or tabu search are suitable to be used within the POPMUSIC framework.

In further research proximity measures for large scale problems with tight time windows will be developed and studied.

Acknowledgments

Financial support from the Oesterreichische Nationalbank (OENB) under grant #11984 and from the Austrian Science Fund (FWF) under grant #P20342-N13 is gratefully acknowledged.

References

[1] Clarke, G., Wright, J.W.: Scheduling of vehicles from a central depot to a number of delivery points. Operations Research 12, 568–581 (1964)
[2] Cordeau, J.-F., Laporte, G., Mercier, A.: A unified tabu search heuristic for the vehicle routing problems with time windows. Journal of the Operation Research Society 52, 928–936 (2001)
[3] Flaberg, T., Hasle, G., Kloster, O., Riise, A.: Towards solving hughe-scale vehicle routing problems for household type applications. In: Workshop presentation in Network Optimization Workshop Saint-Remy de Provence, France (August 2006)
[4] Fleszar, K., Osman, I.H., Hindi, K.S.: A variable neighbourhood search algorithm for the open vehicle routing problem. European Journal of Operational Research (2008) (accepted for publication)
[5] Gillet, B., Miller, L.: A heuristic algorithm for the dispatch problem. Operations Research 22, 340–349 (1974)
[6] Hansen, P., Mladenović, N.: An introduction to variable neighborhood search. In: Voss, S., Martello, S., Osman, I.H., Roucairol, C. (eds.) Meta-Heuristics: Advances and Trends in Local Search Paradigms for Optimization, pp. 433–458. Kluwer Academic Publishers, Dordrecht (1999)

[7] Hemmelmayr, V.C., Doerner, K.D., Hartl, R.F.: A variable neighborhood search heuristic for periodic routing problems. European Journal of Operational Research (2008) (accepted for publication)

[8] Homberger, J., Gehring, H.: A two-phase hybrid metaheuristic for the vehicle routing problem with time windows. European Journal of Operational Research 162, 220–238 (2005)

[9] Kytöjoki, J., Nuortio, T., Bräysy, O., Gendreau, M.: An efficient variable neighborhood search heuristic for very large scale vehicle routing problems. Computers & Operations Research 34, 2743–2757 (2007)

[10] Lin, S.: Computer solutions of the traveling salesman problem. Bell System Technical Journal 44, 2245–2269 (1965)

[11] Mester, D., Bräysy, O.: Active guided evolution strategies for large-scale vehicle routing problems with time windows. Computers & Operations Research 32, 1593–1614 (2005)

[12] Mester, D., Bräysy, O.: Active-guided evolution strategies for large-scale capacitated vehicle routing problems. Computers & Operations Research 34, 2964–2975 (2007)

[13] Ostertag, A., Hartl, R.F., Doerner, K.F., Taillard, E.D., Waelti, P.: Popmusic for a real world large scale vehicle routing problem with time windows. Journal of the Operational Research Society (2008) (accepted for publication)

[14] Polacek, M., Doerner, K., Hartl, R.F., Maniezzo, V.: A variable neigborhood search for the capacitated arc routing problem with intermediate facilities. Journal of Heuristics (2008) (accepted for publication)

[15] Polacek, M., Hartl, R.F., Doerner, K., Reimann, M.: A variable neighborhood search for the multi depot vehicle routing problem with time windows. Journal of Heuristics 10, 613–627 (2004)

[16] Reimann, M., Doerner, K., Hartl, R.F.: D-ants: Saving based ants divide and conquer the vehicle routing problem. Computers & Operations Research 31, 563–591 (2004)

[17] Taillard, E.D., Badeau, P., Gendreau, M., Guertin, F., Potvin, J.Y.: A tabu search heuristic for the vehicle routing problem with soft time windows. Transportation Science 31, 170–186 (1997)

[18] Taillard, E.D., Voss, S.: Popmusic: Partial optimization metaheuristic under special intensification conditions. In: Ribeiro, C., Hansen, P. (eds.) Essays and surveys in metaheuristics, pp. 613–629. Kluwer Academic Publishers, Dordrecht (2001)

A Memetic Algorithm with Population Management (MA|PM) for the Periodic Location-Routing Problem

Caroline Prodhon and Christian Prins

Institut Charles Delaunay, Université de Technologie de Troyes, BP 2060, 10010 Troyes Cedex, France
{christian.prins,caroline.prodhon}@utt.fr

Abstract. A generalization of the well-known Vehicle Routing Problem (VRP) has been developed toward tactical or strategic decision levels of companies but not both. The tactical extension or Periodic VRP (PVRP) plans a set of trips over a multiperiod horizon, subject to frequency constraints. The strategic extension is motivated by interdependent depot location and routing decisions in most distribution systems. Low-quality solutions are obtained if depots are located first, regardless the future routes. In the Location-Routing Problem (LRP), location and routing decisions are simultaneously tackled. The goal here is to combine the PVRP and LRP into an even more realistic problem covering all decision levels: the Periodic LRP or PLRP. An evolutionary algorithm called Memetic Algorithm with Population Management (MA|PM) is proposed to solve large size instances of the PLRP. First, a population is randomly generated. Every individual represents a feasible solution using the same combination of visit days on each customers. The evolution is operated by a memetic mechanism and the offsprings must satisfy a distance test before entering the population. Information about better customer assignment to visit days is collected on the offsprings, and is used to create a new population of solutions. The algorithm stops when a given number of regenerations of the population is reached. The method is evaluated on three sets of instances and solutions are compared to the literature on particular cases such as one-day horizons (LRP) or one depot (PVRP). This metaheuristic outperforms the previous method for the PLRP.

Keywords: Heuristic, Periodic Location-Routing Problem.

1 Introduction

Companies desiring to lower their expenses have to pay attention to their logistics costs. Indeed, the latter often represent a large portion of their budget and involve various decision levels. Among them, depot location and vehicle routing are crucial choices. They are usually tackled separately to reduce the complexity of the global problem. However, researches have shown that this strategy often leads to suboptimal solutions [30]. The Location-Routing Problem (LRP)

M.J. Blesa et al. (Eds.): HM 2008, LNCS 5296, pp. 43–57, 2008.

integrates these two decision levels. In general, the LRP is formulated as a deterministic node routing problem (i.e., customers are located on nodes of the network), but a few authors have studied stochastic cases [15,6] and arc routing versions [11,13]. Nevertheless, as shown in [20], most of the published papers consider either capacitated routes or capacitated depots [2,33] but rarely both except very recently [34,3,25,24,4,26].

Beside the strategic aspect of depot location, a focus on tactical decision such as Vehicle Routing Problems (VRP) leads to consider some extensions. One of these consists in integrating frequency constraints on visited customers over a given multiperiod horizon. The resulting problem is known as periodic VRP or PVRP, introduced in 1984 by Christofides and Beasley [8]. As for the LRP, there exist arc-routing versions of the problem [9,14] but most published papers consider a node routing version. The methods used to solve PVRP are mainly heuristics [8,32,7]. A powerful approach is the tabu search algorithm proposed by Cordeau *et al.* [10]. Very recently, Hemmelmayr *et al.* [12] developed a variable neighborhood search heuristic leading to even better results on average.

As special case, the PVRP can also be viewed as a multidepot VRP (MDVRP). The latter is defined on a single day but instead of visiting the customers from routes assigned to a single depot, the vehicles performs from one of the set of depots. Thus, by considering the routing from each depot as the routing from each period of the horizon, the statement of the MDVRP can be seen as a particular PVRP. In such a case, exact methods are available [16,17,21] and report optimal solutions on instances involving until 80 customers (asymmetric problem).

The LRP and the PVRP have been combined in [29] for the first time into an even more realistic problem: the periodic LRP or PLRP. The objective is to determine the set of depots to open, the combination of service days to assign to customers and the routes originating from each depot for each period of the horizon, in order to minimize the total cost. In [29], the proposed method is a metaheuristic. Each global iteration of the algorithm begins by considering the entire set of customers within a single fictive day to determine a subset of depots to open over the horizon. At this point, statistics on the edges appearing in that LRP solution are recorded. The statistics are used to assign a combination of service days to each customer, with respect to their required service frequency. More precisely, the algorithm tries to gather in a same day customers having a great chance to be successive in a route from a good PLRP solution. Then, the remaining problem can be decomposed into independent MDVRP, one per day. It is solved by the Randomized Extended Clarke and Wright Algorithm proposed in [25]. A local search exchanges customers' combination of service days and the algorithm handles another MDVRP according to the new assignment. This alternance performs until convergence occurs. Then, a new global iteration begins with a diversification on the subset of open depots.

In this paper, the proposed method is a genetic based metaheuristic that tries to take into consideration several decision levels. Each global iteration of the algorithm begins by assigning a fixed combination of service days to each customer, with respect to their required service frequency, for the entire set of individual (solution) from the population. During that global iteration, the evolution is tackled by a memetic algorithm with population management scheme. A local search on the assignment of a combination of visit days to customers allows to record a possible better assignment of service days that would be used in the next global iteration of the method.

The paper is organized as follows. The problem is defined in more details in Section 2. Section 3 describes the framework of the proposed algorithm. The performances of the method are evaluated in Section 4. Some concluding remarks close the paper.

2 Problem Definition

The problem studied in this paper is defined on a horizon H composed of P periods (days) and a complete, weighted and undirected network $G = (V, E, C)$. V is a set of nodes comprising of a subset I of m possible depot locations and a subset $J = V \backslash I$ of n customers. The traveling cost between any two nodes i and j is given by c_{ij}. A capacity W_i and an opening cost O_i are associated with each depot site $i \in I$. Each customer $j \in J$ has to be served a given number of times $s(j)$ during the horizon, and $Comb_j$ is its set of allowed combinations of service days. d_{jlr} is the demand of customer j on the day l of combination $r \in Comb_j$. A set K of N identical vehicles of capacity Q is available over H. A vehicle used at least once from a depot during the horizon incurs a fixed cost F and it may perform one single route per day. The total number of vehicles T_i used at depot i is the maximum number of routes performed from depot i over H. It is a decision variable. Figure 1 illustrates the meaning of T_i.

	Depot 1	Depot 2	Depot 3
Day 1	2	2	1
Day 2	3	2	2
Day 3	1	2	3
Day 4	2	2	2
T_i	3	2	3

Total number of vehicles used: $T_1 + T_2 + T_3 = 8$

Fig. 1. Example of the number of routes performed by depot on each day of the horizon

The following constraints must hold:

- each customer j must be served exclusively on each day l of exactly one of the combination $r \in Comb_j$, by one vehicle and with the amount d_{jlr};
- the sum of the vehicles assigned to depots (T_i) not exceed N;
- each route must begin and end at the same depot within the same day and its total load must not exceed vehicle capacity;
- the total load of the routes assigned to a depot on any day $l \in H$ must fit the capacity of that depot.

The total cost of a route includes the fixed cost F and the costs of traversed edges (variable costs) on each day of the horizon. The objective is to find which subset of depots should be opened, which combination of days should be assigned to each customers and which routes should be performed, to minimize the total cost (fixed costs of depots, plus total cost of the routes).

The PLRP is obviously NP-hard since it reduces to the well-known VRP when $m = 1$ and $|H| = 1$. It is much more combinatorial than the VRP and therefore, due to the size of the targeted instances, a metaheuristic is proposed. It is an evolutionary approach in which successive populations of solutions are tackled by a genetic scheme complemented by a local search (memetic algorithm). Furthermore, a population management based on a distance measure is implemented: a solution is accepted only if its distance to the current population in not smaller than a given threshold with respect to a cost measure. The next section describes the algorithm more in details.

3   MA|PM for the PLRP

The PLRP involves three kinds of decisions. It would be very difficult and time-consuming to search a neighborhood that simultaneously handles these decisions. However, trying to keep a vision the more global as possible on the whole problem is important. That is what is proposed in this paper.

Evolutionary algorithms have been successfully applied to vehicle routing problems, especially the genetic algorithms hybridized with local search, also called memetic algorithms (MA), [18,22]. Recently, Sörensen and Sevaux [31] proposed a new form called MA|PM or memetic algorithm with population management. MA|PM is characterized by a small population Pop of $PopSize$ individuals, the improvement of new solutions by local search, and the replacement of the traditional mutation operator by a distance-based population management technique. Given a threshold Δ, a new solution is accepted only if its distance to Pop is at least Δ. Otherwise, two options are conceivable: either the offspring is mutated until its distance to Pop reaches the threshold [31], or it is simply discarded [27]. These authors described also several dynamic control policies for Δ.

MA|PM has already been applied to the Capacitated Arc Routing Problem [27]. The results indicate that it converges faster than conventional memetic algorithms. Applications to other routing problems have also been developed, leading to very good results as on the production-distribution problem [5] or the LRP [24].

Moreover, its general structure is simpler than other distance-based population metaheuristics such as scatter search or path relinking, and it is quite easy to upgrade an existing MA into an MA|PM.

The choice of MA|PM for the PLRP has been inspired by these promising characteristics. The version studied here corresponds to the second option: children that do not match the threshold are simply discarded. Section 3.1 presents the structure of the chromosomes used, Section 3.2 explains the crossover phase, while Sections 3.3 and 3.4 respectively develop the local search and the population management technique.

3.1   Chromosomes

A good representation of the solution into chromosome is crucial for the achievement of the method. In particular, it is important to design fixed length chromosomes because they are required by most crossovers.

However, encoding information about the depots, the assignments of customers (to service days and to a depot) and the order of deliveries to these customers within a chromosome is not trivial.

That is why we have chosen to set a fixed combination of service days to the customers for each individual from the population. A diversification on this assignment will be tackled later on. Thus, we can use the idea proposed in [24] for the LRP, in which the adopted chromosome encoding comprises a depot status part DS and a customer sequence part CS (see Figure 2).

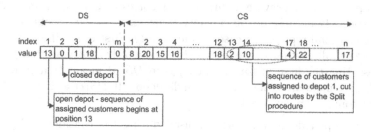

Fig. 2. Representation of an LRP solution as a chromosome

DS (depot status) is a vector of m numbers. $DS(i)$ represents the status of depot i indicating whether it is closed (zero) or opened (non-zero value). If it is opened, $DS(i)$ is the index in CS of the first customer assigned to depot i. CS (customers sequence) contains the concatenation of the lists of customers assigned to the considered service day, without trip delimiters. So, CS is a permutation of customers and has a fixed length.

As we work on a periodic version of the problem, an individual becomes a set of P chromosomes (one for each day of the horizon), taking care that the open depots are the same in each DS vectors.

The fitness $F(S)$ of an individual S is the total cost of the associated PLRP solution. DS is used to deduce the cost of open depots and the list of customers assigned to each depot. Each list can be optimally partitioned into trips using a procedure called Split. This procedure was originally developed by Prins for the VRP [22]. For a list of p customers, it builds an auxiliary graph $G_a = (X, A, Z)$, where X contains $p + 1$ nodes indexed from 0 to p, A contains one arc (i, j), $i < j$, if a trip servicing customers S_{i+1} to S_j (included) is feasible in terms of capacity. The weight z_{ij} of (i, j) is equal to the trip cost. The optimal splitting of the list corresponds to a min-cost path from node 0 to node p in G_a. For the PLRP, we apply Split to the list of customers of each depot and on each day.

3.2   Selection of Parents and Crossover

To generate an individual, the first step is to select its two parents. The first one comes from a binary tournament among the solutions of the population, and the second one from a binary tournament on the whole population except the first selected parent. Then, basically, for two parents A and B, and on each day of the horizon, a one-point crossover is applied to the DS vectors of the two parents and another one to their CS vectors. The crossover for DS works like for binary chromosomes. The one for CS is adapted for permutations. The offspring C receives the sub-sequence of A located before the cutting point. B is then scanned from left to right, starting from the cutting point. The customers not yet in C are copied in CS at the same position as in B to complete the offspring. Once arrived at the end of B, if there are still some vacant positions in CS, they are filled by non-inserted customers.

This operator may provide a chromosome corresponding to an infeasible solution, especially because of the capacity constraints on the depots. Thus, each child is tested and repaired in case of infeasibility. First of all, the repair procedure checks if all customers are assigned to a depot by verifying that the value 1 (representing the first customer in CS) is in DS. It means that in the worst case, all customers are assigned to a single depot. If $DS(i) \neq 1$ for each $i \in I$ three possibilities are explored:

1. the first open depot i found not used during the considered day (depot opened over the horizon but with no route on this day), if any exists, is considered and $DS(i) = 1$;
2. otherwise, a closed depot i is opened and $DS(i) = 1$;
3. if all depots are already opened and used, a depot i is randomly chosen and $DS(i) = 1$.

Then, the procedure looks for depots having a capacity violation. If one depot is found, this means that too many customers are assigned to it. Therefore, the algorithm scans backward the sequence of customers assigned to such a depot and removes them one by one until the capacity constraint of the depot holds. Removed customers are assigned to the first opened depot found having enough remaining capacity. If none exists, its closest depot is not yet opened is used.

3.3   Local Searches for the PLRP

MA|PM works on a small population of high-quality solutions [31]. This quality results from the application of a local search procedure on the offspring. For the PLRP, a local search LS is applied on the solution deduced from the chromosome. It is based on the three following neighborhoods:

- *MOVE*. One customer is shifted from its current position to another position, in the same route or in a different route which may be assigned to the same depot or not, provided capacities are respected.

- *SWAP*. Two customers are exchanged. They may belong to the same route or, if residual capacities allow it, to two distinct routes sharing one common depot or not.
- *OPT*. This is a 2-opt procedure proposed in [25] and in which two non-consecutive edges are removed, either in the same route or in two distinct routes assigned to a common depot or not. When they belong to different routes, there are various ways of reconnecting the trips. If they are from different depots, edges connecting the last customers of the two considered routes to their depot have to be replaced to satisfy the constraint imposing that a route must begin and finish at the same depot. This neighborhood is equivalent in the first case to the well-known 2-OPT move for the TSP [19].

LS executes the first improving move found in the three neighbourhoods (not the best move) and stops when no such move can be found. Note that a depot can only be opened by the crossover and not by these procedures. The assignment to service day will be managed by the diversification phase.

LS must not be called systematically, to avoid a premature convergence and also because it is time-consuming with its $O(n^2)$ neighborhoods on each day of the horizon. In practice, it is applied to the offspring with a fixed probability p_{LS}.

3.4 Population Management

The population management filters the entrance of the offspring in the population thanks to a distance measure. Because of the complexity of a PLRP solution, the distance here is not measured in the solution space, but in the objective space.

In MA|PM, a new solution T may enter the population Pop only if $d_{Pop}(T) \geq \Delta$, where Δ is a given threshold. If $\Delta = 0$, the algorithm behaves like a traditional MA. $\Delta > 0$ ensures distinct solutions in Pop. If Δ takes a large value, most children-solutions are rejected and the MA spends too much time in unproductive iterations. Δ can be dynamically adjusted between such extremes to control population diversity. Different control policies are suggested in [31]. In our MA|PM, Δ is initialized to a rather high value Δ_{max}. If a series of $MaxNbRej$ successive rejections of the offsprings is reached, Δ decreases until accepting a new offspring in the population, but keeping the minimum value of Δ equal to Δ_{min}. Then, Δ remains constant until a new series of successive rejections occurs. If a new best solution is found, Δ is reset to Δ_{max}.

The offspring, if accepted by the population management system described above, replaces the worst chromosome in the current population.

3.5 Diversification

The mechanisms presented until now do not deal with the assignment of the customers to service days. Though, this decision is very important to achieve good results. A technique having a view over the horizon would improve to the solution cost.

Thus, a diversification search intends to reduce the routing cost of each child from a population by finding a new combination of visit days to customers. A move is performed if the best insertion cost of the customers in the days of the new combination is lower than the cost to serve it at its current position. Of course, the capacity constraints must hold to accept the move.

As each individual of a population must have the same assignment to service days, the obtained solution is not included in Pop. However, if the solution is the best one encountered during the evolution of this population, the resulting information about the assignment is recorded.

Then, when a given number $MaxNbNoAdd$ of children has already been discarded by the population management, an effective diversification is made. It refreshes the population by creating new individuals on the basis of assignment to service days recorded during the evolution of the previous population.

3.6 Algorithm

Figure 3 gives an overview of the proposed method. The first assignment of each customer j to a combination from $Comb_j$ is set by gathering within a same day customers having great chance to be successive in a route from the optimal solution. Thus, we build a spanning tree on each period in accordance with the possible combinations of visit days. In order to try to keep a balance over the horizon, one customer is iteratively assigned to each day.

4 Computational Study

4.1 Instances

The proposed method is evaluated on three sets of randomly generated Euclidean instances. The two first ones are made of 30 instances with a set of homogenous capacitated vehicles and a set of possible capacitated depots. Their main characteristics are the followings: number of depots $m \in \{5, 10\}$, number of clients $n \in \{20, 50, 100, 200\}$ vehicle capacity $Q \in \{70, 150\}$ and number of clusters $\beta \in \{0, 2, 3\}$. The case $\beta = 0$ corresponds in fact to a uniform distribution of customers in the Euclidean plane. These instances in which all numbers are integer were generated as follows. For given choices of m, n, Q, and β, the customers' locations are randomly chosen in the Euclidean plane, and the traveling costs c_{ij} correspond to the distances, multiplied by 100 and rounded up to the nearest integer. Each demand follows a uniform distribution in interval $[10, 20]$.

To mimic a working week, *the first set*, especially generated for the PLRP, is made of a 7-day cyclic horizon H with 2 idle days (Saturday and Sunday). The visit frequency of the customers $s(i)$ is between once and three times a week. The allowed set of combinations of visit days $comb(i)$ are given in Table 1 and forbids visits on two consecutive days. The demand d_{jlr} of customer j on each day l depends on the chosen service combination r. It varies with the number of days separating two visits, but it can be pre-calculated. The demand over the

```
 1: BestCost := +∞
 2: NbAcc := 0
 3: NbRej := 0
 4: NbGen := 0
 5: Assign customers to service days by building a tree per day
 6: repeat
 7:    NbGen = NbGen + 1
 8:    Δ := Δ_max
 9:    NbNoAdd := 0
10:    GenPop(Pop) for a given assignment of customers to service days
11:    repeat
12:       Selection(A,B)
13:       Crossover(A,B,C)
14:       if random < p_LS then
15:          LS(C)
16:       end if
17:       Local Search on service days (C)
18:       if cost(C) < BestCost then
19:          BestCost := cost(C)
20:          Δ := Δ_max
21:          BestSoln := C
22:       end if
23:       if cost(C) < BestItCost then
24:          Record the corresponding assignment to service days
25:       end if
26:       if d_P(C) < Δ then
27:          NbRej := NbRej + 1
28:          NbNoAdd := NbNoAdd + 1
29:       else
30:          NbRej:= 0
31:          NbAcc := NbAcc +1
32:          AddToPop(C)
33:          Δ := Δ_max
34:       end if
35:       if NbRej > MaxNbRej then
36:          Δ := Max(Δ_min,Δ * RedDelta)
37:       end if
38:    until NbNoAdd > MaxNbNoAdd
39: until NbGen > MaxNbGen
40: Return (BestSoln)
```

Fig. 3. MA|PM for the Periodic Location-Routing Problem

horizon is divided by the number of service days, leading to an average demand by day. Then, for each service day l of a combination r, d_{jlr} is this average demand multiplied by the number of days since the last service day. For the first day of the combination, the d_{jlr} is the difference of the demand over the horizon minus the demand served on each other day from the combination ($\sum_{k \neq l} d_{jkr}$).

Table 1. Allowed set of combinations of visit days

Frequency	Combinations of visit days
1	Monday
	Tuesday
	Wednesday
	Thursday
	Friday
2	Monday - Wednesday
	Monday - Thursday
	Tuesday - Friday
3	Monday - Wednesday - Friday

The results obtained on these instances are compared with respect to the results proposed for the first time on these instances in [29].

The second set contains LRP instances, created for our previous works [23,25,26] and may be downloaded at [28]. It is reused in this study to compare the performances of proposed method with respect to the best-known results (BKR), obtained when testing various methods with different parameters.

Finally, *the third set* comprises 30 instances for the PVRP available at [1]. The original set has 32 instances, but 2 are discarded by the authors because they contain customers having the same visit frequency but not the same combination of visit days. The horizon is made of P periods and the demand is equal in each service day. The fleet size is limited to k and each vehicle has a capacity Q. The number of customers n ranges from 20 to 417. The traveling costs are equal to the Euclidean distances (not rounded). The performances of our algorithm are compared with the best-known results (BKR) on these instances.

4.2   Implementation, Parameters and Algorithms Compared

The proposed algorithm is coded in Visual C++ and has been tested on a Dell Latitude D420, with an Intel Centrino Duo 1.2 GHz, 1 GB of RAM and running Windows XP.

The following parameters have been selected after a preliminary testing phase, in order to provide the best average solution values: $PopSize = 25 + (n+m)/10$, $MaxNbRej= 5$, $MaxNbNoAdd= (n+m)/3$, $P1 = 0.35$, Δ_{max} corresponds to the average cost of an edge in the complete graph, $\Delta_{min} = \Delta_{max}/10$, $RedDelta = 0.8$ and the maximal number of global regenerations $MaxNbGen= 4|H|$.

4.3   Discussion of Results

In the following tables, each line corresponds to an instance or a subset of instances having the same characteristics. Costs indicate the average value of the objective function on the subset and times T are given in seconds. The columns reported as BKR are the average of the best-known results of the instances. On LRP and PVRP instances, they come from the respective websites [28,1]. On

Table 2. Results on the Periodic Location-Routing Problem

| | | IM | | MAPM | | |
n	m	Cost	CPU	Cost	CPU	Gap
20	5	78882.00	0.63	77969.75	0.35	-1.37
50	5	168027.63	6.55	155608.63	3.01	-7.19
100	5	282062.67	52.45	264635.17	33.90	-6.11
100	10	266575.17	117.72	269634.33	40.48	1.62
200	10	449690.33	950.66	470604.83	754.85	5.34
Average			226.00		166.70	-1.93

PLRP instances, IM column corresponds to the results obtained by the iterative metaheuristic proposed in [29]. The gaps are the deviation in percentage between each method and BKR or IM taken as reference.

First set. Table 2 provides a comparison between the proposed MA|PM and our earlier iterative metaheuristic (IM) detailed in [29] on the 30 PLRP instances from the first set. In IM, each global iteration of the algorithm begins by considering the entire set of customers within a single fictive day to determine a subset of depots to open over the horizon and allocates each customer to a combination of service days. Then, the remaining problem is an MDVRP per day. It is solved by using our Randomized Extended Clarke and Wright Algorithm (RECWA) for the LRP [25].

The results show that the proposed method outperforms the IM, with an improvement of 1.93% of the cost on average, while the CPU time is divided by almost 2 (IM was executed on a Dell PC Optiplex GX260, with a 2.4 GHz Pentium 4, 512 MB of RAM and Windows XP). It is even able to reduce the total cost up to 23% on particular instance in comparison with the already known solutions.

The very good performances are especially noticeable on small and middle size instances. With up to 100 customers and 5 depots, the average improvement is getting around 5.5%. However, when the number of depots increase to 10, the results are not so relevant. A 7% decrease of the cost can be reach with 100 customers, but the gaps becomes mainly in favor to IM on most of the instances.

These observations confirm the importance on the choice of depots and the proposed MA|PM seems to hardly manage that point with respect to IM. Notice that in IM the depot location is the first step of the algorithm while in MA|PM the first decision is made on the allocation of the customers to service days.

Second set. Table 3 gives a comparison on the 30 LRP instances from the second set, between the results from the proposed MA|PM and the best-known solutions (BKR) available on [28], obtained during the testing of various metaheuristics dedicated to the LRP.

The table shows that the proposed metaheuristic is able to provide good solutions even on other problems than the one it is dedicated to. The gap with BKS is small (2.89% on average when the best method on this problem [26], is

Table 3. Results on the Location-Routing Problem

		BKS	MAPM		
n	m	Cost	Cost	CPU	Gap
20	5	45086.75	45167.75	0.06	0.16
50	5	74112.63	76068.38	2.57	2.45
100	5	199164.67	205676.50	42.95	3.19
100	10	238544.00	251136.33	49.15	4.88
200	10	420029.50	432663.00	1019.45	3.01
Average				223.00	2.89

at 0.5%). Some best-known solutions are reached, especially on small instances but even on one with 50 customers.

The CPU time is multiplied by two in comparison with another MA|PM (executed on a Dell PC Optiplex GX260, with a 2.4 GHz Pentium 4, 512 MB of RAM and Windows XP) especially developed for the LRP, that has a gap with the BKR at 1.1%. However, the parameters used in this study are kept unchanged for all problems, but they can be tuned to provide more convenable solutions. Note also that it is able to provide a total cost 6.4% better than the MA|PM for the LRP on an instance with 100 customers and 10 depots.

The gap never rises above 6% except on 3 instances and surprisingly, the number of available depots does not seem to affect a lot the quality of the solution.

Table 4. Results on the Periodic Vehicle Routing Problem

I	n	k	P	Q	BKS Cost	MAPM Cost	CPU	Gap
1	50	3	2	160	524.61	559.39	0.94	6.63
2	50	3	5	160	1322.87	1357.40	4.08	2.61
3	50	1	5	160	524.61	677.18	0.63	29.08
4	75	5	5	140	835.43	921.02	3.09	10.24
5	75	6	10	140	2027.99	2135.52	14.84	5.30
6	75	1	10	140	836.37	890.97	4.31	6.53
7	100	4	2	200	826.14	848.02	14.73	2.65
8	100	5	5	200	2034.15	2124.36	55.09	4.43
9	100	1	8	200	826.14	856.78	5.88	3.71
10	100	4	5	200	1595.84	1727.25	39.78	8.23
12	163	3	5	140	1195.88	1271.75	107.70	6.34
13	417	9	7	2000	3511.62	—	—	—
14-16	20-56	2	4	20-40	1897.56	1897.56	2.48	0.00
17-20	40-184	4	4	20-60	4486.68	4530.70	232.33	1.81
21-22	60-114	6	4	20-30	3227.58	3382.28	41.80	4.60
24-26	51	3	6	20	3753.31	3932.90	2.86	4.80
27-29	102	6	6	20	22600.18	24721.01	24.42	9.37
30-32	153	9	6	20	77891.89	89822.04	98.08	15.35
Average							56.81	6.58

Third set. Table 4 provides a comparison on the PLRP instances from the third set, between the results obtained by the proposed MA|PM and the best-known solutions (BKR) available from [1]. When no solution appears in the table, the algorithm does not succeed in finding a solution compatible with the fleet size.

Gaps are larger on average on PVRP instances showing how the periodic aspect is hard to deal with. However, the results remain good with a gap around 5% in comparison with the best methods especially designed for the PVRP [10,12]. The cost are also at 5% larger on average than BKR on instances with up to 5 periods. Note also that even if the PVRP has not been as studied as VRP, the first works on this topics appeared 30 years ago and many papers have been published on this problem since that time. Furthermore, the proposed method is not especially designed for PVRP but for much more combinatorial ones and the required CPU time is almost the same as for the other algorithms while taking decisions from several levels.

5   Conclusion

In this paper, a new metaheuristic for the Location Routing Problem (LRP) with both capacitated depots and vehicles is presented. The method is an evolutionary algorithm called memetic algorithm with population management (MA|PM). It is a genetic method hybridized with local search techniques and a distance measure to control the population, plus a diversification search which manages the period aspect. The method has been tested on three sets of small, medium and large scale instances, and compared to other heuristics on various kind of instances such as one-day horizons (LRP) or one depot (PVRP). The solutions obtained show that this algorithm is able to find good solutions and outperform the previous algorithm dedicated to the PLRP.

References

1. http://neo.lcc.uma.es/radi-aeb/WebVRP/ (2007)
2. Albareda-Sambola, M., Díaz, J.A., Fernández, E.: A compact model and tight bounds for a combined location-routing problem. Computers and Operations Research 32, 407–428 (2005)
3. Barreto, S.S.: Análise e Modelização de Problemas de localização-distribuição [Analysis and modelling of location-routing problems]. PhD thesis, University of Aveiro, campus universitário de Santiago, 3810-193 Aveiro, Portugal (in Portuguese) (October 2004)
4. Belenguer, J.M., Benavent, E., Prins, C., Prodhon, C., Wolfler-Calvo, R.: A cutting plane method for the capacitated location-routing problem. In: Odysseus 2006 - Third International Workshop on Freight Transportation and Logistics, Altea, Espagne, pp. 50–52 (May 2006)
5. Boudia, M., Louly, M.A.O., Prins, C.: A memetic algorithm with population management for a production-distribution problem. In: Dolgui, A., Morel, G., Pereira, C.E. (eds.) Preprints of the 12th IFAC Symposium on Information Control Problems in Manufacturing INCOM 2006, Saint Etienne-France, vol. 3, pp. 541–546 (May 2006)

6. Chan, Y., Carter, W.B., Burnes, M.D.: A multiple-depot, multiple-vehicle, location-routing problem with stochastically processed demands. Computers and Operations Research 28, 803–826 (2001)
7. Chao, M., Golden, B.L., Wasil, E.: An improved heuristic for the periodic vehicle routing problem. Networks 26, 25–44 (1995)
8. Christofides, N., Beasley, J.E.: The period routing problem. Networks 14, 237–256 (1984)
9. Chu, F., Labadi, N., Prins, C.: A scatter search for the periodic capacitated arc routing problem. European Journal of Operational Research 169, 586–605 (2006)
10. Cordeau, J.F., Gendreau, M., Laporte, G.: A tabu search heuristic for periodic and multi-depot vehicle routing problems. Networks 30, 105–119 (1997)
11. Ghiani, G., Laporte, G.: Location-arc routing problems. OPSEARCH 38, 151–159 (2001)
12. Hemmelmayr, V.C., Doerner, K.F., Hartl, R.F.: A variable neighborhood search heuristic for periodic routing problems. European Journal of Operational Research (in press)
13. Labadi, N.: Problèmes tactiques et stratégiques en tournées sur arcs. PhD thesis, University of Technology of Troyes (2003)
14. Lacomme, P., Prins, C., Ramdane-Chérif, W.: Evolutionary algorithms for periodic arc routing problems. European Journal of Operational Research 165(2), 535–553 (2005)
15. Laporte, G., Louveaux, F., Mercure, H.: Models and exact solutions for a class of stochastic location-routing problems. European Journal of Operational Research 39, 71–78 (1989)
16. Laporte, G., Norbert, Y., Arpin, D.: Optimal solutions to capacitated multi-depot vehicle routing problems. Congressus Numerantium 44, 283–292 (1984)
17. Laporte, G., Norbert, Y., Taillefer, S.: Solving a family of multi-depot vehicle routing and location-routing problems. Transportation Science 22, 161–167 (1988)
18. Lima, C.M.R.R., Goldbarg, M.C., Goldbarg, E.F.G.: A memetic algorithm for heterogeneous fleet vehicle routing problem. Electronic Notes in Discrete Mathematics 18, 171–176 (2004)
19. Lin, S., Kernighan, B.W.: An effective heuristic algorithm for the traveling salesman problem. Operations Research 21, 498–516 (1973)
20. Min, H., Jayaraman, V., Srivastava, R.: Combined location-routing problems: a synthesis and future research directions. European Journal of Operational Research 108, 1–15 (1998)
21. Mingozzi, A., Valletta, A.: An exact algorithm for period and multi-depot vehicle routing problems. In: Odysseus 2003 - Second International Workshop on Freight Transportation and Logistics, Palermo, Italy (May 2003)
22. Prins, C.: A simple and effective evolutionary algorithm for the vehicle routing problem. Computers and Operations Research 31, 1985–2002 (2004)
23. Prins, C., Prodhon, C., Wolfler-Calvo, R.: Nouveaux algorithmes pour le problème de localisation et routage sous contraintes de capacité. In: Dolgui, A., Dauzère-Pérès, S. (eds.) MOSIM 2004, Nantes, France, vol. 2, pp. 1115–1122. Lavoisier (September 2004)
24. Prins, C., Prodhon, C., Wolfler-Calvo, R.: A memetic algorithm with population management (MA|PM) for the capacitated location-routing problem. In: Gottlieb, J., Raidl, G.R. (eds.) EvoCOP 2006. LNCS, vol. 3906, pp. 183–194. Springer, Heidelberg (2006)

25. Prins, C., Prodhon, C., Wolfler-Calvo, R.: Solving the capacitated location-routing problem by a GRASP complemented by a learning process and a path relinking. 4OR - A Quarterly Journal of Operations Research 4, 221–238 (2006)
26. Prins, C., Prodhon, C., Wolfler-Calvo, R.: Solving the capacitated location-routing problem by a cooperative lagrangean relaxation-granular tabu search heuristic. Transportation Science 41(4), 470–483 (2007)
27. Prins, C., Sevaux, M., Sörensen, K.: A genetic algorithm with population management (GA | PM) for the carp. In: Tristan V (5th Triennal Symposium on Transportation Analysis), Le Gosier, Guadeloupe, June 13-18 (2004)
28. Prodhon, C. (2007), http://prodhonc.free.fr/homepage
29. Prodhon, C.: An iterative metaheurtistic for the periodic location-routing problem. In: GOR 2007, Saarbrücken (September 2007)
30. Salhi, S., Rand, G.K.: The effect of ignoring routes when locating depots. European Journal of Operational Research 39, 150–156 (1989)
31. Sörensen, K., Sevaux, M.: MA | PM: memetic algorithms with population management. Computers and Operations Research 33, 1214–1225 (2006)
32. Tan, C.C.R., Beasley, J.E.: A heuristic algorithm for the periodic vehicle routing problem. OMEGA International Journal of Management Science 12(5), 497–504 (1984)
33. Tuzun, D., Burke, L.I.: A two-phase tabu search approach to the location routing problem. European Journal of Operational Research 116, 87–99 (1999)
34. Wu, T.H., Low, C., Bai, J.W.: Heuristic solutions to multi-depot location-routing problems. Computers and Operations Research 29, 1393–1415 (2002)

Memetic Algorithm for Intense Local Search Methods Using Local Search Chains

Daniel Molina[1], Manuel Lozano[2],
C. García-Martínez[3], and Francisco Herrera[2]

[1] University of Cádiz, Department of Computer Languages and Systems, Cádiz
daniel.molina@uca.es
[2] University of Granada, Department of Computer Science and Artificial Inteligence,
Granada
lozano@decsai.ugr.es, herrera@decsai.ugr.es
[3] University of Córdoba, Department of Computing and Numerical Analysis, Córdoba
cgarcia@uco.es

Abstract. This contribution presents a new memetic algorithm for continuous optimization problems, which is specially designed for applying intense local search methods. These local search methods make use of explicit strategy parameters to guide the search, and adapt these parameters with the purpose of producing more effective solutions. They may achieve accurate results, at the cost of requiring high intensity, making more difficult their application into a memetic algorithm. Our memetic algorithm approach assigns to each individual a local search intensity that depends on its features, by chaining different local search applications. With this technique of search chains, at each stage the local search operator may continue the operation of a previous invocation, starting from the final configuration reached by this one. The proposed memetic algorithm integrates the CMA-ES algorithm as their local search operator. We compare our proposal with other memetic algorithms and evolutionary algorithms for continuous optimization, showing that it presents a clear superiority over the compared algorithms.

1   Introduction

It is now well established that hybridisation of *evolutionary algorithms* (EAs) with other techniques can greatly improve the efficiency of search [1,2]. EAs that have been hybridised with local search techniques are often called *memetic algorithms* (MAs) [3,4,5]. One commonly used formulation of MAs improvement the new member of the population using a local search (LS) method, with the aim of exploiting the best search regions gathered during the global sampling done by the EA. That allow to design MAs for continuous optimisation (MACOs) that obtain very accuracy solutions in these type of problems [6].

For function optimization problems in continuous search spaces, an important difficulty must be addressed: solutions of high precision must be obtained by the solvers. MAs comprising efficient local improvement processes on continuous

M.J. Blesa et al. (Eds.): HM 2008, LNCS 5296, pp. 58–71, 2008.

domains (continuous LS methods) have been presented to deal with this problem [6]. In this paper, they will be named MACOs (MAs for continuous optimization problems).

Most well-known continuous LS algorithms make use of explicit *strategy parameters* (e.g., *step sizes*) to guide the search. Generally, they adapt these parameters with the purpose of increasing the likelihood of producing more effective solutions. Because of their explicit parameter adaptation, these algorithms may require a substantial number of evaluations to achieve adequate styles of traversal of solution space to follow certain paths leading to precise final solutions. We call these local search *intense local search methods*. For this behaviour, the usual hybridization model is not adequate enough to these intense continuous LS algorithms, because the total function evaluations invested by the LS operator may become too high, hindering to obtain profitable synergetic effects between the EA and the LS algorithm.

In this contribution, we present a MACO model specially designed to incorporate intense continuous LS methods as LS operators. Our proposal applies the local search with an adaptive intensity, exploiting with higher intensity the most promising individuals. To adapt the LS intensity, our proposal can apply the LS operator several times over the same individual, using a fixed LS intensity, creating *LS chains*.

With this technique of *LS chains*, an individual resulting from a LS invocation can later become the initial point of a subsequent LS application, using the final strategy parameter values achieved by the former as its initial ones. In this way, the continuous LS method may *adaptively* fit its strategy parameters to the particular features of these zones. In our study, we use CMA-ES [7] as intense continuous LS algorithm, which stands out as an excellent local searcher.

This contribution is set up as follows. In Section 2, we present different aspects of *intense* local search and it is described the CMA-ES algorithm. In Section 3, we present the concept of LS chain and how it can be applied to improve the integration with *intense* continuous local searches. In Section 4, we present an experimental study to compare our proposal with other algorithms proposed in the literature. Finally, in Section 5, we provide the main conclusions of this work.

2   Intense Continuous Local Search Algorithms and CMA-ES

In his pioneer work on MACOs, Hart [6] demonstrated that the choice of continuous LS algorithm affects the performance of MACOs significantly on a variety of benchmark problems with diverse properties.

Most well-known continuous LS algorithms make use of explicit *strategy parameters* (e.g., *step sizes*) to guide the search. Generally, they adapt their parameters, in such a way that the moves being made may be of varying sizes, depending on the success of previous steps, with the purpose of increasing the likelihood of producing more effective solutions. Due to their explicit parameter adaptation, these continuous LS algorithms may require high LS intensity values to adapt their strategy parameters to the local topography of the search areas being refined. They are called intense continuous LS algorithms.

The integration of intense continuous LS algorithms into MACOs arises as a research area particularly attractive, because, nowadays, there are advanced intense continuous LS algorithms that stand out as formidable local searchers. The *covariance matrix adaptation evolution strategy* (CMA-ES) [7,8] is one of them. CMA-ES was originally introduced to improve the LS performance of evolution strategies. Even though CMA-ES even reveals competitive global search performances [9], it has exhibited effective abilities for the local tuning of solutions. At the *2005 Congress of Evolutionary Computation*, a *multi-start LS metaheuristics* using these method, called L-CMA-ES [10], was one of the winners of the real-parameter optimization competition [11,12]. Thus, investigating the behaviour of CMA-ES as LS component for MACOs deserves much attention.

In CMA-ES, not only is the step size of the mutation operator adjusted at each generation, but so too is the step direction in the multidimensional problem space, by a covariance matrix whose elements are updated as the search proceeds. In this work, we use the (μ_W, λ) CMA-ES model. For every generation, this algorithm generates a population of λ offspring by sampling a multivariate normal distribution:

$$x_i \sim N\left(m, \sigma^2 C\right) = m + \sigma N_i(0, C) \text{ for } i = 1, \cdots, \lambda,$$

Where the mean vector m represents the favourite solution at present, the so-called step-size σ controls the step length, and the covariance matrix C determines the shape of the distribution ellipsoid. Then, the μ best offspring are used to recalculate the mean vector, σ and m and the covariance matrix C, following equations that may be found in [7] and [9]. The default strategy parameters are given in [9]. Only the initial m and σ parameters have to be set depending on the problem.

It can be interpreted any evolution strategy that uses intermediate recombination as a LS strategy [7]. Thus, since CMA-ES is extremely good at detecting and exploiting local structure, it turns out to be a particularly reliable and highly competitive EA for local optimization [10]. Also, it can be described as an intense continuous LS algorithm, because, as noted by Auger, Schoenauer and Vanhaecke [13]: *"CMA-ES may require a substantial number of time steps for the adaptation of the covariance matrix"*.

3   MACOs Based on Local Search Chains

Due to the potential of the intense continuous LS algorithms, it becomes interesting to build MACO models with them. These MACOs should be specifically designed to accomplish two essential aims:

- The intense continuous LS algorithm should be provided with sufficient LS intensity to make their correct operation possible.
- The MACO should ensure that a profitable synergy between the continuous LS algorithm and the EA is possible.

In this section, we propose a MACO approach conceived to attain these two objectives. It is a steady-state MA model that employs the concept of *LS chain* to adjust the LS intensity assigned to the intense continuous LS method. In particular, our MACO handles LS chains, throughout the evolution, with the objective of allowing the continuous LS algorithm to act more intensely in the most promising areas represented in the EA population. In this way, the continuous LS method may adaptively fit its strategy parameters to the particular features of these zones.

In Section 3.1, we introduce the foundations of steady-state MAs. In Section 3.2, we present the concept of *LS chain*. In Section 3.3, we propose a MACO approach that handles LS chains with the objective of make good use of intense continuous LS methods as LS operators. Finally, in Section 3.4, we present an instance of our MACO model that uses CMA-ES as continuous LS operator.

3.1 Steady-State MAs

In *steady-state* GAs [14] usually only one or two offspring are produced in each generation. Parents are selected to produce offspring and then a decision is made as to which individuals in the population to select for deletion in order to make room for the new offspring. Steady-state GAs are *overlapping* systems because parents and offspring compete for survival.

Although steady-state GAs are less common than generational GAs, Land recommended their use for the design of *steady-state MAs* (steady-state GAs plus LS) because they may be more stable (as the best solutions do not get replaced until the newly generated solutions become superior) and they allow the results of LS to be maintained in the population. So, steady-state MAs integrate global and local search more tightly than generational MAs [15]. This interleaving of the global and local search phases allows the two to influence each other.

3.2 Local Search Chains

In steady-state MAs, individuals resulting from the LS invocation may be for a long time latter selected to be replaced. At this point, we propose *to chain* an LS algorithm invocation and the next one as follows:

> *The final configuration reached by the former (strategy parameter values, internal variables, etc.) is used as initial configuration for the next application.*

In this way, the LS algorithm may continue under the same conditions achieved when the LS operation was previously halted, providing an *uninterrupted connection between successive LS invocations*, i.e., forming a *LS chain*. Figure 1 shows an example of LS chain formed by a LS algorithm with only one associated strategy parameter, p.

Two important aspects that were taken into account for the management of LS chains are:

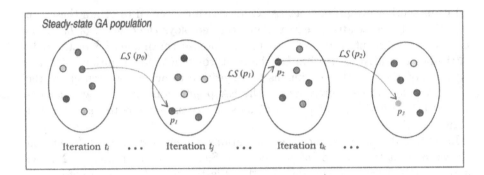

Fig. 1. Example of LS chain. p_x is the value for the strategy value, p_{i+1} is the final parameter value reached when it started with a value of p_i, and p_0 is its default value

- Every time the LS algorithm is applied to refine a particular chromosome, it is applied a fixed LS intensity, that will be called *LS intensity stretch* (I_{str}). In this way, a LS chain formed throughout n_{app} LS applications and started from solution s_0 will return the same solution as the application of the continuous LS algorithm to s_0 employing $n_{app} \cdot I_{str}$ fitness function evaluations.
- After the LS operation, the parameters that define the current state of the LS processing are stored along with the reached final individual (in the steady-state GA population). When this individual is latter selected to be improved, the initial values for the parameters of the LS algorithm will be directly available.

In this work, we argue that a promising approach to *adapt* the LS intensity assigned to intense continuous LS algorithms is using MACOs that allow LS chain to grow throughout the evolution depending on the quality of the search regions being visited, with the aim of acting more intensely in the most promising areas. In this way, the real LS intensity assigned to the continuous LS algorithm may be adaptively determined throughout the run and depends on two crucial choices:

- The way the solutions are selected to apply the LS operator to them.
- The replacement scheme used by the steady-state GA.

The designer of the steady-state GA is responsible for the second election, whereas the first one should be undertaken during the design of the MACO scheme.

3.3   A MACO Model That Handles Local Search Chains

In this section, we propose a MACO model (see Figure 2) with the following main features:

1. It is a steady-state MA model.
2. It ensures that a fixed and predetermined local/global search ratio is always kept. With this policy, we easily stabilise this ratio, which has a strong influence on the final MACO behaviour, avoiding an excessive exploitation.

3. It favours the enlargement of those LS chains that are showing promising fitness improvements in the best current search areas represented in the steady-state GA population. In addition, it encourages the activation of innovative LS chains with the aim of refining unexploited zones, whenever the current best ones may not offer profitability. The criterion to choose the individuals that should undergo LS is specifically designed to manage the LS chains in this way (Steps 3 and 4).

---

1. Generate the **initial population**.
2. Perform the **steady-state GA** throughout n_{frec} evaluations.
3. Build the set S_{LS} with those individuals that **potentially may be refined by LS**.
4. Pick **the best individual** in S_{LS} (Let's c_{LS} to be this individual).
5. if c_{LS} belongs to an **existing LS chain** then
6.   Initialise the LS operator with the **LS state stored** together with c_{LS}.
7. else
8.   Initialise the LS operator with the **default LS state**.
9. Apply the LS algorithm to c_{LS} with an LS intensity of I_{str} (Let's c_{LS}^r to be the resulting individual).
10. Replace c_{LS} by c_{LS}^r in the **steady-state GA population**.
11. Store the **final LS state** along with c_{LS}^r.
12. If (*not termination-condition*) go to step 2.

Fig. 2. Pseudocode algorithm for the proposed MACO model

The proposed MACO scheme defines the following relation between the steady-state GA and the intense continuous LS method (Step 2): *every n_{frec} number of evaluations of the steady-state GA, apply the continuous LS algorithm to a selected chromosome, c_{LS}, in the steady-state GA population.* Since we assume a fixed $\frac{L}{G}$ ratio, $r_{L/G}$, n_{frec} may be calculated using the equation $n_{frec} = I_{str} \frac{1-r_{L/G}}{r_{L/G}}$, where n_{str} is the LS intensity stretch (Section 3.2). We recall that $r_{L/G}$ is defined as the percentaje of evaluations spent doing local search from the total assigned to the algorithm's run.

The following mechanism is performed to select c_{LS} (Steps 3 and 4):

1. Build the set of individuals in the steady-state GA population, S_{LS} that fulfils:
   (a) They have never been optimised by the intense continuous LS algorithm, or
   (b) They previously underwent LS, obtaining a fitness function improvement greater than δ_{LS}^{min} (a parameter of our algorithm).

With this mechanism, when the steady-state GA finds a new best so far individual, it will be refined immediately. In addition, the best performing individual in the steady-state GA population will always undergo LS whenever the fitness improvement obtained by a previous LS application to this individual is greater than the δ_{LS}^{min} threshold.

3.4   Memetic Algorithm with LS Chaining and CMA-ES

In this section, we build an instance of the proposed MACO model (Figure 2), which applies CMA-ES (Section 2) as intense continuous LS algorithm. It will be called MA-LSCh-CMA. Next, we list the main features of this algorithm:

Steady-state GA. It is a real-coded steady-state GA [16] specifically designed to promote high population diversity levels by means of the combination of the BLX-α crossover operator with a high value for its associated parameter ($\alpha = 0.5$) and the *negative assortative mating* strategy [17], in combination with the *replacement strategy*. Diversity is favoured as well by means of the BGA mutation operator [18]. This combination of selection parents and replacement strategy let achieve an adequate trade-off between exploration and exploitation, th In the MA literature, keeping population diversity while using LS together with an EA is always an issue to be addressed, either implicitly or explicitly [19,20].

CMA-ES as Continuous LS algorithm. MA-LSCh-CMA follows the MACO approach, presented in Section 3.3, to handle LS chains, with the objective of tuning the intensity of CMA-ES, which is employed as intense continuous LS algorithm (Section 2). The application of CMA-ES for refining an individual, C_i, is carried out following the next guidelines:

- C_i becomes the initial mean of distribution (m).
- The initial σ value is half the distance of C_i to its nearest individual in the steady-state GA population (this value allows an effective exploration around C_i).

CMA-ES will work as local searcher consuming I_{str} fitness function evaluations. Then, the resulting solution will be introduced in the steady-state GA population along with the current value of the covariance matrix, the mean of the distribution, the step-size, and the variables used to guide the adaptation of these parameters (B, BD, D, p_c and p_σ). Latter, when CMA-ES is applied to this inserted solution, these values will be recovered to proceed with a new CMA-ES application. When CMA-ES is performed on solutions that do not belong to an existing chain, default values, given in [9], are assumed for the remaining strategy parameters.

Parameter setting. For the experiments, MA-LSCh-CMA applies BLX-α with $\alpha = 0.5$. The population size is 60 individuals and the probability of updating a chromosome by mutation is 0.125. The n_{ass} parameter associated with the negative assortative mating is set to 3. The value of the $\frac{L}{G}$ ratio, $r_{L/G}$, was set to 0.5, which represents an well-balanced choice. Finally, a value of 1e-8 was assigned to the δ_{LS}^{min} threshold.

4 Experiments

We have carried out different experiments to assess the performance of MA-LSCh-CMA. In order to do this, in this section, we detail the test functions and the experimental setup and statistical methods that were used for this experimental study.

This section is structure as following: In Section 4.1 it is presented the test functions applied for the experiments. In Section 4.2 there are presented the of the experiments. In Section 4.3 analyses the influence of the LS intensity stretch in our proposal. In Section 4.4 it is studied the convenience of the LS chaining. In Section 4.4, we compare our proposal with other modern metaheuristics with the L-CMA-ES and in section 4.5 with the DEahcSPX algorithms. Finally, in Section 4.6, they are shown the numerical results (average error) obtained by each one of the algorithms considered in this Section.

4.1 Test Functions

The test suite that we have used for different experiments consists of 20 benchmark functions chosen from the set designed for the *special session on real parameter optimisation organised in the 2005 IEEE congress on evolutionary computation* (CEC2005). We have considered only the multimodal functions (F6-F25) from the CEC2005 test suite; because we are particularly interested in analysing its behaviour with complicated test functions. It is possible to consult in [11] the complete description of the functions. Also, we have considered the dimension 30, because we want to focus our study on the most dificult problems.

4.2 Experimental Setup and Statistical Analysis

The experiments have been done following the instructions indicated in the document associated to the competition. The main characteristics are:

- Each algorithm is run 25 times for each test function, and the error average of the best individual of the population is computed.
- The study has been made with dimensions $D = 30$.
- The maximum number of fitness evaluations that we allowed for each algorithm to minimise the error was $10,000 \cdot N$, where N is the dimension of the problem.
- Each run stops either when the error obtained is less than 10^{-8}, or when the maximal number of evaluations is achieved.

We have carried out the experimental study of MA-LSCh-CMA and the other algorithms following these guidelines in order to make possible its comparison with the results of all the other algorithms involved in the competition (their results are available in the proceedings of the congress).

To analyse the results we have chosen to use *non-parametric tests* because it has been proven that in this benchmark the parametric tests cannot be applied with security [21]. In particular, we have considered two alternative methods based on non parametric tests to analyse the experimental results, previously applied in comparisons of EAs [21]:

- Application of the *Iman and Davenport's* test and the *Holm's* method as post-hoc procedure. The first test is used to see whether there are significant statistical differences among the algorithms in certain group. If differences are detected, then Holm's test is employed to compare the best algorithm (control algorithm) against the remaining ones.
- Utilization of the *Wilcoxon* matched-pairs signed-ranks test. Using this test, the results of two algorithms may be directed compared.

In [21] these statistical tests are explained in detail.

4.3   Influence of the LS Intensity Stretch

In our first empirical study, we investigate the influence of I_{str} on the performance of MA-LSCh-CMA. In particular, we analyse the behaviour of this algorithm when different values for this parameter are considered ($I_{str} = 100$, 500, and 1000).

First, it is applied the *Iman-Davenport* tests at the 5% level, Table 1 shows the results.

Table 1. Results of the Iman-Davenport's test with different I_{str} values

Iman-Davenport value	Critical value	Sig. differences
1,17	2,77	No

From Table 1, we may extract an important conclusion: MA-LSCh-CMA exhibits a low sensitivity degree to the value selected for I_{str}. We have chosen a particular value for I_{str}, in order to allow the incoming study of our proposal and the comparison with other EA models to be easily understandable.

Figure 3 shows the average rankings obtained by the MA-LSCh-CMA instances with different I_{str} values on the test functions for the different dimensions. The height of each column is proportional to the ranking, *the lower a column is, the better its associated algorithm is*. In order to study these results, we may see that $I_{str} = 500$ is the best choice, so it is the selected value.

4.4   Studying the Behaviour of the Proposed MACO Model

In this section, we have performed two different experiments, in order to investigate the behaviour of the proposed MACO model.

Comparison with a Standard MACO. First, we want to check if the LS chain offers an improvement over a standard MACO using a CMA-ES as its

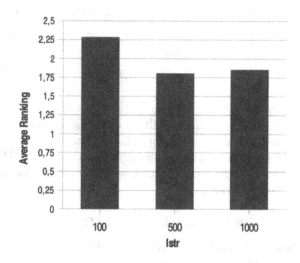

Fig. 3. Rankings obtained by MA-LSCh-CMA instances with different I_{str} values

LS method, which will be denoted as S-MACO. The basic difference between S-MACO and MA-LSCh-CMA is that the former always selects the best performing individual in the steady-state GA population as the one to be improved by CMA-ES, which starts from default values for its strategy parameters and consumes I_{LS} evaluations. It has been tested S-MACO with three different values for I_{LS} were investigated: 100, 500, and 1000, obtained that 1000 is the best value for I_{LS}. We should point out that S-MACO fits the $\frac{L}{G}$ ratio to 0.5, such as MA-LSCh-CMA does.

So, we have compared MA-LSCh-CMA ($I_{str} = 500$) with S-MACO with $I_{LS} = 1000$, using Wilcoxon's test. Table 2 summarizes the results of this procedure, where the values of $R+$ and $R-$ (associated to MA-LSCh-CMA) of the test are specified (the lowest ones, which correspond to the best results, are highlighted in bold face), together with the critical values.

We clearly notice that MA-LSCh-CMA obtains better results than S-MACO (the $R-$ value is lower than the $R+$ one). But in addition, the statistical test indicates that these improvements are statistically significant (because the $R-$ value is lower than the critical value).

Comparison with a Restart Local Search Algorithm In this section, we carry out the comparison of MA-LSCh-CMA with a restart CMA-ES, called

Table 2. S-MACO versus MA-LSCh-CMA using Wilcoxon's test (p-value $= 0.05$)

$R+$ (S-MACO)	$R-$ (MA-LSCh-CMA)	Critical value	Sig. differences?
168	**42**	52	Yes

Table 3. L-CMA-ES versus MA-LSCh-CMA (Wilcoxon's test with p-value = 0.05)

R+ (L-CMA-ES)	R− (MA-LSCh-CMA)	Critical value	Sig. differences?
165	45	52	Yes

L-CMA-ES [10] because both algorithms invoke CMA-ES instances that specifically emphasise the local refinement abilities of this algorithm. Table 3 has the results of the comparison of these two algorithms using the Wilcoxon's test.

MA-LSCh-CMA exhibits overall better performance than L-CMA-ES, therefore, the work of the proposed hybridization method outperforms the one of the pure restart local search strategy.

4.5 Comparison with State-of-the-Art MACOs

In a recent publication, it has been presented a MACO model, called DEahcSPX [22], that combines differential evolution with a quick continuous LS method. DEahcSPX was compared with other MACO instances proposed in the literature, and they found that their proposal was superior to the majority of them. Thus, we assume that DEahcSPX is currency the most outstanding representative of the state-of-the-art MACOs.

In this section, we undertake the comparative analysis among DEahcSPX and MA-LSCh-CMA using Wilcoxon's test. Table 4 contains the results of this statistical test.

The results of MA-LSCh-CMA show higher quality than the ones of DEahcSPX. In addition, the superiority is statistically significant. Thus our proposal has turned out to be very competitive with state-of-the-art MACOs.

Then, we may highlight that MA-LSCh-CMA arises as one of the most prominent algorithm for global optimization over continuous spaces, by two important search processes simultaneously:

- The steady-state GA induces a scattered search promoting population diversity.
- The proliferation of long LS chains in the best regions becomes suitable to obtain adequate *accuracy* levels, letting also to search in alternative regions.

4.6 Results of Experiments

We present in this section the results of our proposal and the differents algorithms used into the comparisons. That allow to compare them with other algorithms

Table 4. DEahcSPX versus MA-LSCh-CMA (Wilcoxon's test with p-value = 0.05)

R+ (DEahcSPX)	R− (MA-LSCh-CMA)	Critical value	Sig. differences?
169,5	40,5	52	Yes

Table 5. Average Errors by each algorithms in the benchmark applied

Test Functions	MA-LSCh CMA	S-MACO	DEahcSPX
F6	1.191003e+1	2.732782e+1	**1.000000e-9**
F7	**8.871392e-4**	2.067364e-3	1.163264e-3
F8	**2.027016e+1**	2.086726e+1	2.094711e+1
F9	7.827714e-9	8.374473e-9	1.000000e-9
F10	**1.838684e+1**	7.243991e+1	9.449920e+1
F11	**4.350834e+0**	9.017085e+0	2.921885e+1
F12	**7.690185e+2**	1.462644e+3	2.956616e+4
F13	2.344814e+0	**2.282783e+0**	2.365826e+0
F14	**1.268192e+1**	1.253313e+1	1.279216e+1
F15	**3.080000e+2**	3.160001e+2	3.506300e+2
F16	1.363134e+2	1.719942e+2	**1.294508e+2**
F17	**1.345630e+2**	1.427101e+2	2.048724e+2
F18	**8.156512e+2**	8.265035e+2	9.060900e+2
F19	**8.163714e+2**	8.237708e+2	9.061617e+2
F20	**8.157765e+2**	8.284801e+2	9.065054e+2
F21	5.120000e+2	5.120000e+2	**5.000000e+2**
F22	5.258481e+2	**5.043677e+2**	9.120960e+2
F23	**5.341643e+2**	5.341645e+2	5.341641e+2
F24	2.000000e+2	2.000000e+2	2.000000e+2
F25	2.108472e+2	**2.092819e+2**	2.105413e+2

using the same benchmark functions. Table 5 shows for each algorithm used its average error with the experimental setup indicated into Section 4.2. It has been remarked in bold type the lower average error for each function.

5 Conclusions

This work presents a new hybridization model specially designed to integrate intense continuous LS methods that need a high intensity. In the proposal model, the continuous LS algorithm is applied with higher intensity in the most promising solutions. It is proposed a MACO algorithm called MA-LSCh-CMA that employs the CMA-ES algorithm using the previous hybridization model.

The proposed MA-LSCh-CMA has been compared, following guidelines recommended for the *CEC 2005 special session on real-parameter optimization*, with other state-of-the-art EAs for continuous optimizations. Our proposal present significant improvements other them.

Other important conclusion is that the new hybridization model opens the design of new MACOs using efficiently a category of local search methods, intense continuous LS methods, that until now could not be easily integrated for requiring a high intensity. The design of new MACOs for other intense LS algorithms using the concept of *LS chains* will be studied as future work.

Acknowledgments

Authors thank Nasimul Noman and Hitoshi Iba for letting us use their DEahc-SPX code, and N. Hansen for kindly providing his implementation of CMA-ES in Matlab.

References

1. Davis, L.: Handbook of Genetic Algorithms. Van Nostrand Reinhold, New York (1991)
2. Goldberg, D.E., Voessner, S.: Optimizing global-local search hybrids. In: Banzhaf, W., et al. (eds.) Proceedings of the Genetic and Evolutionary Computation Conference 1999, pp. 220–228. Morgan Kaufmann, San Mateo (1999)
3. Moscato, P.: On evolution, search, optimization, genetic algorithms and martial arts: Towards memetic algorithms. Technical report, Technical Report Caltech Concurrent Computation Program Report 826, Caltech, Pasadena, California (1989)
4. Moscato, P.: Memetic algorithms: a short introduction, pp. 219–234. McGraw-Hill, London (1999)
5. Merz, P.: Memetic Algorithms for Combinatorial Optimization Problems: Fitness Landscapes and Effective Search Strategies. PhD thesis, University of Siegen, Germany
6. Hart, W.: Adaptive Global Optimization With Local Search. PhD thesis, Univ. California, San Diego, CA (1994)
7. Hansen, N., Ostermeier, A.: Adapting Arbitrary Normal Mutation Distributions in Evolution Strategies: The Covariance Matrix Adaptation. In: Proceeding of the IEEE International Conference on Evolutionary Computation (ICEC 1996), pp. 312–317 (1996)
8. Hansen, N., Müller, S., Koumoutsakos, P.: Reducing the Time Complexity of the Derandomized Evolution Strategy with Covariance Matrix Adaptation (CMA-ES). Evolutionary Computation 1(11), 1–18 (2003)
9. Hansen, N., Kern, S.: Evaluating the CMA Evolution Strategy on Multimodal Test Functions. In: Yao, X., Burke, E.K., Lozano, J.A., Smith, J., Merelo-Guervós, J.J., Bullinaria, J.A., Rowe, J.E., Tiño, P., Kabán, A., Schwefel, H.-P. (eds.) PPSN 2004. LNCS, vol. 3242, pp. 282–291. Springer, Heidelberg (2004)
10. Auger, A., Hansen, N.: Performance Evaluation of an Advanced Local Search Evolutionary Algorithm. In: 2005 IEEE Congress on Evolutionary Computation, pp. 1777–1784 (2005)
11. Suganthan, P., Hansen, N., Liang, J., Deb, K., Chen, Y., Auger, A., Tiwari, S.: Problem definitions and evaluation criteria for the CEC 2005 special session on real parameter optimization. Technical report, Nanyang Technical University (2005)
12. Hansen, N.: Compilation of Results on the CEC Benchmark Function Set. In: 2005 IEEE Congress on Evolutionary Computation (2005)
13. Auger, A., Schoenauer, M., Vanhaecke, N.: LS-CMAES: a second-order algorithm for covariance matrix adaptation. In: Yao, X., Burke, E.K., Lozano, J.A., Smith, J., Merelo-Guervós, J.J., Bullinaria, J.A., Rowe, J.E., Tiño, P., Kabán, A., Schwefel, H.-P. (eds.) PPSN 2004. LNCS, vol. 3242, pp. 182–191. Springer, Heidelberg (2004)
14. Whitley, D.: The GENITOR Algorithm and Selection Pressure: Why Rank-Based Allocation of Reproductive Trials is Best. In: Proc. of the Third Int. Conf. on Genetic Algorithms, pp. 116–121 (1989)
15. Land, M.S.: Evolutionary Algorithms with Local Search for Combinational Optimization. PhD thesis, Univ. California, San Diego, CA (1998)
16. Herrera, F., Lozano, M., Verdegay, J.L.: Tackling Real-coded Genetic Algorithms: Operators and Tools for the Behavioral Analysis. Artificial Intelligence Reviews 12(4), 265–319 (1998)

17. Fernandes, C., Rosa, A.: A Study of non-Random Matching and Varying Population Size in Genetic Algorithm using a Royal Road Function. In: Proc. of the 2001 Congress on Evolutionary Computation, pp. 60–66 (2001)
18. Mülenbein, H., Schlierkamp-Voosen, D.: Predictive Models for the Breeding Genetic Algorithm in Continuous Parameter Optimization. Evolutionary Computation 1, 25–49 (1993)
19. Lozano, M., Herrera, F., Krasnogor, N., Molina, D.: Real-coded Memetic Algorithms with Crossover Hill-climbing. Evolutionary Computation 12(2), 273–302 (2004)
20. Tang, J., Lim, M., Ong, Y.: Diversity-adaptive parallel memetic algorithm for solving large scale combinatorial optimization problems. Soft Computing 11(9), 873–888 (2007)
21. García, S., Molina, D., Lozano, M., Herrera, F.: A study on the use of non-parametric tests for analyzing the evolutionary algorithms' behaviour: A case study on the cec 2005 special session on real parameter optimization. Journal of Heuristics (in press, 2008)
22. Noman, N., Iba, H.: Accelerating differential evolution using an adaptive local search. In: IEEE Transactions on evolutionary Computation (in press, 2008)

Incremental Particle Swarm-Guided Local Search for Continuous Optimization

Marco A. Montes de Oca[1], Ken Van den Enden[2], and Thomas Stützle[1]

[1] IRIDIA, CoDE, Université Libre de Bruxelles, Brussels, Belgium
{mmontes,stuetzle}@ulb.ac.be
[2] Vrije Universiteit Brussel, Brussels, Belgium
kvdenden@vub.ac.be

Abstract. We present an algorithm that is inspired by theoretical and empirical results in social learning and swarm intelligence research. The algorithm is based on a framework that we call *incremental social learning*. In practical terms, the algorithm is a hybrid between a local search procedure and a particle swarm optimization algorithm with growing population size. The local search procedure provides rapid convergence to good solutions while the particle swarm algorithm enables a comprehensive exploration of the search space. We provide experimental evidence that shows that the algorithm can find good solutions very rapidly without compromising its global search capabilities.

1 Introduction

Many algorithms that have been successfully used for solving optimization problems can be thought of as being a collection of agents that interact with each other and with the environment. In fact, this mental imagery has been used to design some of them. Some examples are genetic algorithms [1], ant colony optimization algorithms [2], and particle swarm optimization algorithms [3]. Following this same approach, we present an algorithm that is inspired by results in social learning and swarm intelligence research. The algorithm is based on a multiagent learning framework that we call *incremental social learning* [4]. The multiagent scenario used for the design of the proposed algorithm is the following: A growing population of agents that can learn both individually and socially explores its environment in order to maximize the group's well-being. The strategy used by the agents is to use their social learning skills when they become part of the population and when learning individually is either too costly or deemed unproductive.

Technically speaking, the algorithm presented in this paper is a hybrid between the well-known Powell's direction set method [5] and a particle swarm optimization algorithm with growing population size [4]. Powell's method is used as an agent's individual learning mechanism. If the local topography of the objective function (from an agent's perspective) allows it, this mechanism improves candidate solutions without any information exchange between agents. The particle swarm velocity- and position-update rules, which allow the global

M.J. Blesa et al. (Eds.): HM 2008, LNCS 5296, pp. 72–86, 2008.

Algorithm 1. Incremental social learning

/* Initialization */
$t \leftarrow 0$
Initialize environment $\mathbf{E}^t$
Initialize primogenial population of agents $\mathbf{X}^t$

/* Main loop */
while Stopping criteria not met **do**
  if Agent addition schedule or criterion is not met **then**
    $\mathbf{X}^{t+1} \leftarrow$ ilearn$(\mathbf{X}^t, \mathbf{E}^t)$ /* Individual learning */
  else
    Create new agent a_{new}
    slearn$(a_{new}, \mathbf{X}^t)$ /* Social learning */
    $\mathbf{X}^{t+1} \leftarrow \mathbf{X}^t \cup \{a_{new}\}$
  end if
  $\mathbf{E}^{t+1} \leftarrow$ update$(\mathbf{E}^t)$ /* Update environment */
  $t \leftarrow t + 1$
end while

exploration of the search space, play the role of the agents' social learning mechanism. Together, these components form an incremental particle swarm-guided local search algorithm for solving continuous optimization problems (Section 3).

We compare the performance of the proposed algorithm with that of other four algorithms (Section 4). The proposed algorithm exhibits the fast convergence of Powell's method with good global search capabilities (Section 5).

2   Incremental Social Learning

In previous work, we introduced a multiagent learning framework, called *incremental social learning*, that combines elements of social and individual learning in order to speed up learning and to allow scalability [4]. Social learning is a term that refers to the class of mechanisms that allow the transmission of knowledge between individuals without the use of genetic material [6,7]. Individual learning, on the other hand, is a process that lets an individual acquire knowledge about its environment by interacting directly with it and without any social influence [8]. Applying social learning ideas to the multiagent learning problem is appealing because learning from others effectively gives agents a shortcut to adaptive information that otherwise would be expensive to acquire [9]. The framework is called incremental because it is based on a growing population of agents that adapts incrementally to the final multiagent environment. When a new agent is added to the population, it learns socially from more experienced agents and learns individually once it is part of it. This strategy is inspired by the fact that, in nature, newborn individuals are particularly favored by social learning because it allows them to learn many skills very rapidly from the adult individuals that surround them [10]. The algorithmic structure of the incremental social learning framework is outlined in Algorithm 1.

At the beginning, the environment and the primogenial population of agents are initialized. An agent addition schedule is used to control the rate at which agents are added to the environment. If no agents are to be added, the agents in the current population learn individually by directly interacting with the environment. If a new agent is scheduled to be part of the population, it learns socially from a subset of the agents in the population before it is added. The environment is then updated (if needed) and the cycle is repeated until a stopping criterion is satisfied.

The learning process starts with a small number of agents because this reduces the interference caused by the co-existence of multiple learning agents. Agents are added according to a problem-dependent schedule in order to create time delays that allow agents that are already part of the population to learn from the environment and to have something to teach to newcomers. When a new agent learns socially, it saves itself the effort required to learn individually what others already have. Incrementally growing the population size is also useful for allocating the minimum number of agents required to solve a particular problem.

3   Incremental Particle Swarm-Guided Local Search

The incremental social learning framework, although conceived for tackling multiagent learning problems, can also be used for designing population-based optimization algorithms. This is because, as we said in Section 1, the multiagent metaphor is usually useful for thinking about how this kind of algorithms work. In [4], a particle swarm optimization algorithm with growing population size (IPSO) was designed as an instantiation of the incremental social learning framework. In IPSO, individual learning is not implemented; instead, the algorithm comprises vertical and horizontal social learning mechanisms. Vertical social learning is a pattern of knowledge transmission between individuals of different generations while horizontal social learning happens between individuals of the same generation [11]. In IPSO, vertical social learning occurs when a new particle is added to the swarm and horizontal social learning is used instead of individual learning. IPSO exhibits a good solution quality vs. time trade-off since it finds solutions of at least the same quality than a particle swarm optimization algorithm with constant population size in fewer objective function evaluations.

Here we present an extension of IPSO that includes an individual learning mechanism. Individual learning is implemented as a local search procedure to simulate a particle's self-improvement process. The addition of an individual learning mechanism is justified not only from a practical point of view but also from a theoretical one. Indeed, since social learning is cheaper than individual learning, it is usually assumed to be always advantageous [9]. However, it has been argued that relying only on socially acquired knowledge is not always the best strategy [12]; instead, individuals should devote some of their time and energy to learn individually or to innovate [9].

As IPSO, the proposed algorithm is based on the particle swarm optimization algorithm in which particles (i.e., potential solutions to an optimization problem) move in the search space with a certain velocity. Each particle is attracted toward its own previous best position (with respect to an objective function) and toward the best position found by the particles in its neighborhood. Neighborhood relations are usually given in advance through a population topology which can be defined by a graph $G = \{V, E\}$, where each vertex in V corresponds to a particle in the swarm and each edge in E establishes a neighbor relation between a pair of particles. The velocity and position updates of a particle i over dimension j are as follows

$$v_{i,j}^{t+1} = \chi \cdot [v_{i,j}^t + \varphi_1 \cdot U_1 \cdot (p_{i,j}^t - x_{i,j}^t) + \varphi_2 \cdot U_2 \cdot (l_{i,j}^t - x_{i,j}^t)], \qquad (1)$$

and

$$x_{i,j}^{t+1} = x_{i,j}^t + v_{i,j}^{t+1}, \qquad (2)$$

where $v_{i,j}^t$ and $x_{i,j}^t$ are the particle's velocity and position at time step t respectively, $p_{i,j}^t$ is the particle's best position so far, $l_{i,j}^t$ is the best position found by the particle's neighbors, φ_1 and φ_2 are two parameters, U_1 and U_2 are two uniformly distributed random numbers in the range $[0, 1)$, and χ is a constriction factor that is used in order to avoid an "explosion" of the particles' velocity. Clerc and Kennedy [13] found the relation $\chi = 2k/ \left| 2 - \varphi - \sqrt{\varphi^2 - 4\varphi} \right|$, where $k \in [0, 1]$, and $\varphi = \varphi_1 + \varphi_2 > 4$, to compute it.

The local search procedure employed in the proposed algorithm is the well-known Powell's direction set method [5] using Brent's technique [14] as the auxiliary line minimization algorithm. The decision of using Powell's method as the local search procedure is based on performance considerations. The proposed algorithm calls repeatedly the local search procedure and thus an efficient one was needed. In this sense, Powell's method quadratic convergence can be very advantageous if the objective function is locally quadratic [15].

The last component of the proposed algorithm is the vertical social learning mechanism which is implemented as a rule that moves the new particle's previous best position from its initial random location in the search space to one that is closer to the previous best position of a particle that serves as a "model" to imitate. The rule is applied in a component-wise fashion as follows

$$p'_{new,j} = p_{new,j} + U \cdot (p_{model,j} - p_{new,j}), \qquad (3)$$

where $p'_{new,j}$ is the new particle's updated previous best position, $p_{new,j}$ is the new particle's original previous best position, $p_{model,j}$ is the model's previous best position and U is a uniformly distributed random number in $[0, 1)$.

The combination of the local search procedure and the particle swarm optimization algorithm with growing population size, as sketched in Algorithm 2, produces an incremental particle swarm-guided local search algorithm whose operation is illustrated in Figure 1.

The proposed algorithm is designed for taking advantage of features that the objective function may have. For example, if the objective function is convex and separable, it should be optimized in a single local search run without the

---

Algorithm 2. Incremental Particle Swarm-Guided Local Search
___
Input: Objective function f and maximum number of iterations t_{max}

/* Initialization */
Create initial particle p_1 and add it to the set of particles $\mathcal{P}$ which is initially empty
Initialize position vector $p_1.x$ to random values within the search range
Initialize velocity vector $p_1.v$ to zero
Set $p_1.pb = p_1.x$

/* Main Loop */
Set $t = 0$ and $k = 1$
repeat
  /* Individual learning */
  for $i = 1$ to k **do**
    Improve $p_i.pb$ through a local search procedure
  end for

  /* Horizontal social learning */
  for $i = 1$ to k **do**
    Move $p_i.x$ using Eqs. 1 and 2
    if $f(p_i.x)$ is better than $f(p_i.pb)$ **then**
      Set $p_i.pb = p_i.x$
    end if
  end for

  /* Population growth and vertical social learning */
  if Agent addition criterion is met **then**
    Create particle p_{k+1} and add it to the set of particles $\mathcal{P}$
    Initialize position vector $p_{k+1}.x$ using Eq. 3
    Initialize velocity vector $p_{k+1}.v$ to zero
    Set $p_{k+1}.pb = p_{k+1}.x$
    Set $k = k + 1$
  end if
  Set $t = t + 1$
  Set $sol = \underset{p_i \in \mathcal{P}}{\mathrm{argmin}}\, f(p_i.pb)$
until $f(sol)$ is good enough or $t = t_{max}$

---

need of using the exploration capabilities provided by the particle swarm algorithm. Similarly, if the objective function has a few local optima, starting with only one particle ensures the minimal use of function evaluations for the exploration of the space. However, the objective function may also have features for which local search and small initial populations are ineffective. For example, if an objective function had large plateaus, the local search procedure would return with no solution improvement after having wasted several function evaluations. Under these circumstances, the algorithm would start making progress (thanks to the exploration capabilities of the particle swarm component) until a critical population size is reached.

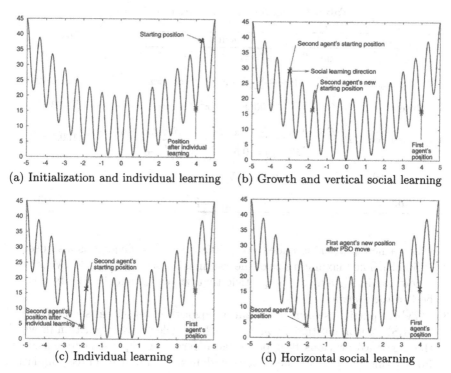

(a) Initialization and individual learning (b) Growth and vertical social learning

(c) Individual learning   (d) Horizontal social learning

Fig. 1. The incremental particle swarm-guided local search algorithm at work. In Figure (a), a particle is randomly placed in the search space. Immediately after, the first round of individual learning (local search) is run, moving the particle to, or close to, a local minimum. This is followed by a horizontal learning round (a normal PSO move) that does not have any effect in a swarm of one particle (the particles' velocity are initialized to zero). In Figure (b), the size of the population is increased. The vertical social learning rule (Eq. 3) is used to place the new particle. In Figure (c), the second iteration of the algorithm begins by running a round of individual learning. In Figure (d), through a round of horizontal learning, particles can move and discover more promising areas of the search space. The algorithm continues intertwining individual and social learning procedures until a stopping criterion is satisfied.

4   Experimental Setup

The performance of the incremental particle swarm-guided local search algorithm (labeled IPSOLS) is compared to that of the following algorithms:

1. A traditional particle swarm optimization algorithm with constant population size (labeled PSO). Three population sizes of 10, 100 and 1000 particles are used.
2. An incremental particle swarm optimization algorithm as described in Section 3 (labeled IPSO). In [4], IPSO proved to work well with a fast agent addition schedule. Accordingly, we use a fast particle addition schedule in

Table 1. Benchmark optimization problems

Name	Definition	Range
Ackley	$-20e^{-0.2\sqrt{\frac{1}{n}\sum_{i=1}^{n}x_i^2}} - e^{\frac{1}{n}\sum_{i=1}^{n}\cos(2\pi x_i)} + 20 + e$	$[-32,32]^n$
Rastrigin	$10n + \sum_{i=1}^{n}(x_i^2 - 10\cos(2\pi x_i))$	$[-5.12,5.12]^n$
Rosenbrock	$\sum_{i=1}^{n-1}[100(x_{i+1} - x_i^2)^2 + (x_i - 1)^2]$	$[-30,30]^n$
Expanded Schaffer	$ES(\boldsymbol{x}) = \sum_{i=1}^{n-1} S(x_i, x_{i+1}) + S(x_n, x_1)$, where $S(x,y) = 0.5 + \frac{\sin^2(\sqrt{x^2+y^2})-0.5}{(1+0.001(x^2+y^2))^2}$	$[-100,100]^n$
Schwefel	$418.9829n + \sum_{i=1}^{n} -x_i\sin(\sqrt{|x_i|})$	$[-500,500]^n$
Sphere	$\sum_{i=1}^{n} x_i^2$	$[-100,100]^n$

which a new particle is added every iteration (this schedule is also used in IPSOLS). The maximum size of the swarm is set to 1000 particles. Whenever a new particle is added, the particle that serves as model to imitate is the best of the swarm.

3. A hybrid particle swarm optimization algorithm with local search (labeled PSOLS). It is a constant population size particle swarm algorithm in which the particles' previous best positions undergo an improvement phase (via Powell's method) before the velocity update rule is applied. Three population sizes of 10, 100 and 1000 particles are used.
4. A random restart local search algorithm (labeled RLS). Every time the local search procedure (Powell's method) converges, it is restarted from a newly generated random solution. The best solution found so far is considered to be the output of the algorithm.

All particle swarm-based algorithms (PSO, IPSO, PSOLS and IPSOLS) are run with two population topologies: a fully connected topology, in which each particle is a neighbor to all others including itself, and the so-called ring topology, in which each particle has two neighbors apart from itself. In the incremental algorithms, the new particle is randomly placed within the topological structure. Other parameter settings are $\varphi_1 = \varphi_2 = 2.05$ and $\chi = 0.7298$.

Powell's method has also a number of parameters to be defined. The so-called tolerance, used to stop the procedure once a very small difference between two solutions is detected, is set to 0.01. In case such a difference is not found, the procedure is stopped after a maximum number of iterations. In our experiments, this parameter is set to 10. Other settings were explored but no significant differences in the results were found. Finally, different step sizes (used for the line minimization procedure) were explored. The values tried for this parameter were 0.1, 1, 10, 20, and 33% of the length of the corresponding search range.

A set of six benchmark functions are used in our experiments. Their mathematical formulation and search ranges are listed in Table 1. In all cases, we used their 100-dimensional instantiations (i.e., $n = 100$).

Our data is based on 100 independent runs of up to 10^6 function evaluations each. In each run, all benchmark functions (except Schwefel's) were randomly shifted within the specified range. Particles or trial points outside the search range were forced to stay within by putting them on the boundary.

5   Results

The distribution of the solution quality obtained by the compared algorithms after 10^6 function evaluations is shown in Figure 2. These results correspond to the case in which all particle swarm-based algorithms use a ring topology and the local search step size is equal to 20% the length of the search range. In [16], the reader can find the complete set of results, including those from the statistical significance tests performed on the algorithms' output data[1]. The algorithms without a local search component are placed on the left side of each plot. This arrangement reveals that the algorithms with a local search component find solutions of equal or better quality than the algorithms without it. The use of a local search component also makes the algorithms more robust to changes in the population topology, that is, the algorithms with a local search component do not show a significant change in their performance if a different topology is used. Only PSO and IPSO are affected by a change in the population topology. For example, the PSO algorithm with 10 particles finds better solutions with the ring topology than with the fully connected topology while the opposite happens if larger populations are used. In any case, regardless of the topology used, the algorithms with local search outperform those without it.

IPSOLS (at the center of all plots) performs at least as well as the other particle swarm-based algorithms with local search. In three problems (Expanded Schaffer, Schwefel, and Rastrigin), IPSOLS obtains better solutions than all other algorithms. With Ackley's function, IPSOLS, PSOLS and RLS obtain comparable results. With Rosenbrock's function, IPSOLS and PSOLS with a population of 10 particles obtain the best results. It is important to note that even after 10^5 function evaluations, the only algorithm that consistently finds high-quality solutions to all our benchmark problems is IPSOLS.

Figure 3 shows the development over time of the median solution quality obtained by the compared algorithms. In order to make the plots more readable, we consider only the algorithms that obtained the best results after 10^6 function evaluations. These plots show more clearly the effects of the incremental and socially guided use of local search procedures. As expected, IPSOLS exhibits the same performance as RLS during the first 10^4 function evaluations. This happens because IPSOLS starts, just as RLS does, performing a local search from a random initial solution. Once the first local search procedure stops making progress, IPSOLS and RLS start differentiating. The beneficial effects of relying on socially acquired information are specially clear in the results obtained for Ackley, Rastrigin, Expanded Schaffer, and Schwefel's problems.

---

[1] Unless otherwise noted, all our statements are supported by statistical evidence.

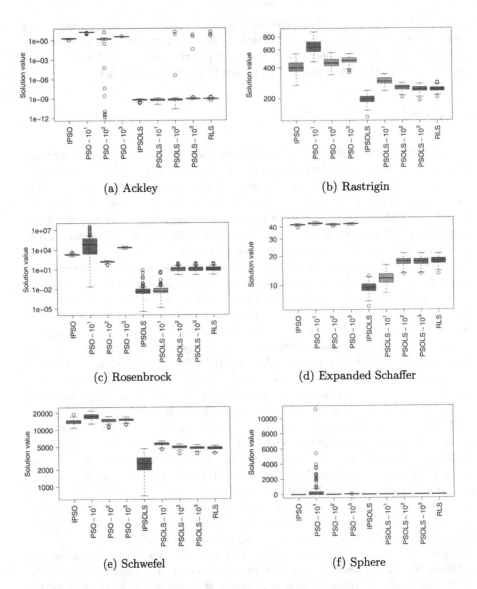

Fig. 2. Boxplots showing the distribution of the solution quality obtained by the compared algorithms after 10^6 function evaluations. These results correspond to the case in which all particle swarm-based algorithms used a ring topology. The numbers next to the labels indicate the population size used.

The results obtained while solving the Sphere problem show clearly that starting with one single particle is beneficial if the objective function is separable and convex. In this case, as can be seen in the plots, IPSOLS and RLS have exactly the same performance. This is so because a single run of the local search procedure is able to find the optimal solution with no extra algorithmic overhead.

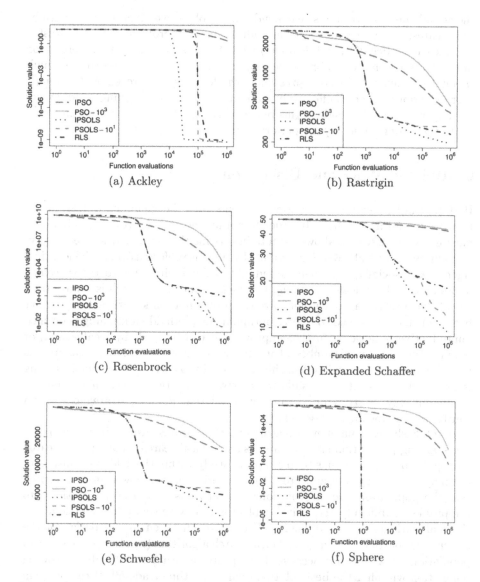

Fig. 3. Development of the solution quality over time (median values). These results correspond to the case in which all particle swarm-based algorithms used a ring topology. The numbers next to the labels indicate the population size used.

The difference between IPSOLS and RLS is that IPSOLS is able to adapt and cope with the difficulties posed by the features of other objective functions.

All the algorithms that contain Powell's method as an algorithmic component are affected by the step size of the subsidiary line minimization algorithm (in our case, Brent's algorithm). Figure 4 shows the median solution quality development over time for different step sizes using RLS. As can be seen in the plots, the actual effect of the step size depends on the objective function. In our study, the greatest differences are seen on multimodal test problems (Ackley, Rastrigin, Expanded Schaffer, and Schwefel) while there is practically no difference on unimodal functions (Rosenbrock and Sphere). This result suggests that, on multimodal problems, large step sizes allow the local search procedure to "jump" over local optima, which effectively increases the chance of finding good quality local optima. The downside of this behavior is that if the step size parameter is set to the wrong value, a poor performance can be obtained.

6   Related Work and Discussion

IPSOLS can be seen as a population of agents with the capability of learning by themselves and from others. In order to obtain the greatest reward, individuals must use a strategy that allows them to intertwine their two modes of learning in a productive way. In IPSOLS, individual learning is preferred over social learning. Agents learn socially only when individual learning is either too costly or deemed unproductive. Practically speaking, this means that local search is always tried first. If local search alone cannot solve the problem at hand satisfactorily, either because it has converged to a local optimum (i.e., individual learning is deemed unproductive because no further improvement can be made by local search) or because the maximum number of iterations of the local search procedure has been reached (i.e., individual learning is deemed too costly), then social learning is used. Of course, it is possible to intertwine the two modes of learning in a different way. A study on the utility of different learning strategies seems a fruitful avenue for future research.

IPSOLS shares a number of features with other algorithms. In particular, a time-varying population size and a subsidiary local search algorithm. Research on population (re)sizing has been a topic of study within the field of evolutionary computation for many years. From that experience, it is now usually accepted that the population size in evolutionary algorithms should be proportional to the problem's difficulty [17]. The problem is that we usually know little about the real-world problem's difficulty. The approach taken in the design of the incremental particle swarm-guided local search algorithm (i.e., to start with small populations) makes sense because if the problem is not so difficult, a growing population will offer the best solution quality vs. time trade-off. If the problem is difficult, there will be a time penalty to pay for reaching very high quality solutions; however, acceptable solutions may be found early in a run.

Practically all resizing strategies consider the possibility of reducing the size of the population during an algorithm's run (see e.g. [18,19,20,21,22,23]). An exception to this approach is the work of Auger and Hansen [24] in which the population size of a CMA-ES algorithm is doubled each time it is restarted. The feature that IPSOLS shares with Auger and Hansen's approach is that no population reduction mechanism is implemented; however, IPSOLS does not use restarts and has a slower population growth rate. Nevertheless, it may be that the performance of IPSOLS is improved by decreasing the population size from

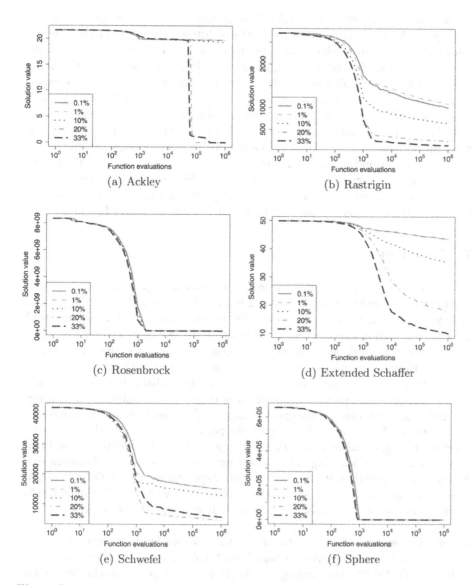

Fig. 4. Development of the solution quality over time (median values) for different local search step sizes using RLS

time to time. This is because decreasing the population size may save many function evaluations. In terms of solution quality, however, it is not expected that this may impact negatively on the quality of the solutions found since the choice of removal would be done on converged particles (particles very close to each other with very small velocity and already in a local minimum).

The idea of combining local search techniques with particle swarm optimization algorithms comes from the observation that particles are attracted by their previous best positions. It is usually thought that the better the attractors are, the higher the chances to find even better solutions. The goal of most hybrid algorithms is thus to accelerate the placement of the particles' previous best positions in good spots. For example, Chen et al. [25] combined a particle swarm algorithm with a hill-climbing local search procedure; Gimmler et al [26] experiment with Nelder and Mead's simplex method as well as with Powell's method; Das et al. [27] also use Nelder and Mead's simplex method and propose the inclusion of an estimate of the local gradient into the particles' velocity update rule. Petalas et al. [28] report experiments with several local search-particle swarm combination schemes. All these previous approaches try to enhance the performance of the particle swarm algorithm; IPSOLS, on the contrary, guides a local search procedure in order to better explore the search space. By doing it incrementally, IPSOLS can exploit the features of the objective function.

7   Conclusions

A hybrid algorithm for solving continuous optimization problems, called incremental particle swarm-guided local search, has been presented. The algorithm has two main components: (i) a particle swarm optimization algorithm with growing population size and (ii) a subsidiary local search procedure that operates on the particles' previous best positions. We presented the experimental evidence that allows us to conclude that the algorithm exhibits a very good exploitation behavior that does not compromise its global search capabilities.

The design of the proposed algorithm is inspired by results in social learning and swarm intelligence research. Using a multiagent systems metaphor, the algorithm can be thought of as a growing population of agents that can learn both socially and individually. The strategy used by the agents to intertwine these two modes of learning is the following: when a naive agent becomes part of the main population, it learns from more experienced agents. Once it is part of the main population, an agent learns socially only when learning individually is considered to be too costly or when doing it is deemed unproductive.

Acknowledgments. Marco A. Montes de Oca is funded by the Programme Alβan, the European Union Programme of High Level Scholarships for Latin America, scholarship No. E05D054889MX, and by the *SWARMANOID* project funded by the Future and Emerging Technologies programme (IST-FET) of the European Commission (grant IST-022888). Thomas Stützle acknowledges support from the F.R.S-FNRS of the French Community of Belgium of which he is a Research Associate.

References

1. Goldberg, D.E.: Genetic Algorithms in Search, Optimization, and Machine Learning. Addison-Wesley, Reading (1989)
2. Dorigo, M., Stützle, T.: Ant Colony Optimization. Bradford Books. MIT Press, Cambridge (2004)

3. Kennedy, J., Eberhart, R.: Particle swarm optimization. In: Proceedings of IEEE International Conference on Neural Networks, pp. 1942–1948. IEEE Press, Piscataway, NJ (1995)
4. Montes de Oca, M.A., Stützle, T.: Towards incremental social learning in optimization and multiagent systems. In: Workshop on Evolutionary Computation and Multiagent Systems Simulation of the Genetic and Evolutionary Computation Conference (GECCO 2008), pp. 1939–1944. ACM Press, New York (2008)
5. Powell, M.J.D.: An efficient method for finding the minimum of a function of several variables without calculating derivatives. The Computer Journal 7(2), 155–162 (1964)
6. Nehaniv, C.L., Dautenhahn, K. (eds.): Imitation and Social Learning in Robots, Humans and Animals: Behavioral, Social and Communicative Dimensions. Cambridge University Press, Cambridge (2007)
7. Curran, D., O'Riordan, C.: Increasing population diversity through cultural learning. Adaptive Behavior 14(4), 315–338 (2006)
8. Aoki, K., Wakano, Y., Feldman, M.W.: The emergence of social learning in a temporally changing environment: A theoretical model. Current Anthropology 46(2), 334–340 (2005)
9. Laland, K.N.: Social learning strategies. Learning & Behavior 32(1), 4–14 (2004)
10. Galef Jr., B.G., Laland, K.N.: Social learning in animals: Empirical studies and theoretical models. BioScience 55(6), 489–499 (2005)
11. Cavalli-Sforza, L.L., Feldman, M.W.: Cultural Transmission and Evolution. A Quantitative Approach. Princeton University Press, Princeton (1981)
12. Giraldeau, L.A., Valone, T.J., Templeton, J.J.: Potential disadvantages of using socially acquired information. Philosophical Transactions of the Royal Society of London. Series B: Biological Sciences 357, 1559–1566 (2002)
13. Clerc, M., Kennedy, J.: The particle swarm–explosion, stability, and convergence in a multidimensional complex space. IEEE Transactions on Evolutionary Computation 6(1), 58–73 (2002)
14. Brent, R.P.: Algorithms for Minimization Without Derivatives. Prentice-Hall, Englewood Cliffs (1973)
15. Press, W.H., Teukolsky, S.A., Vetterling, W.T., Flannery, B.P.: Numerical Recipes in C. The Art of Scientific Computing, 2nd edn. Cambridge University Press, Cambridge (1992)
16. Montes de Oca, M.A., Van den Enden, K., Stützle, T.: Incremental particle swarm-guided local search for continuous optimization: Complete results, (2008), http://iridia.ulb.ac.be/supp/IridiaSupp2008-013/
17. Lobo, F.G., Lima, C.F.: Adaptive Population Sizing Schemes in Genetic Algorithms. In: Parameter Setting in Evolutionary Algorithms. 2007 of Studies in Computational Intelligence, vol. 54, pp. 185–204. Springer, Berlin (2007)
18. Arabas, J., Michalewicz, Z., Mulawka, J.J.: GAVaPS – A genetic algorithm with varying population size. In: Proceedings of the IEEE Conference on Evolutionary Computation, pp. 73–78. IEEE Press, Piscataway, NJ (1994)
19. Harik, G.R., Lobo, F.G.: A parameter-less genetic algorithm. In: Banzhaf, W., et al. (eds.) Proceedings of the Genetic and Evolutionary Computation Conference (GECCO 1999), pp. 258–265. Morgan Kaufmann, San Francisco (1999)
20. Bäck, T., Eiben, A.E., van der Vaart, N.A.L.: An empirical study on GAs "without parameters". In: Deb, K., Rudolph, G., Lutton, E., Merelo, J.J., Schoenauer, M., Schwefel, H.-P., Yao, X. (eds.) PPSN 2000. LNCS, vol. 1917, pp. 315–324. Springer, Heidelberg (2000)

21. Eiben, A.E., Marchiori, E., Valkó, V.A.: Evolutionary algorithms with on-the-fly population size adjustment. In: Yao, X., Burke, E.K., Lozano, J.A., Smith, J., Merelo-Guervós, J.J., Bullinaria, J.A., Rowe, J.E., Tiňo, P., Kabán, A., Schwefel, H.-P. (eds.) PPSN 2004. LNCS, vol. 3242, pp. 41–50. Springer, Heidelberg (2004)
22. Fernandes, C., Rosa, A.: Self-regulated population size in evolutionary algorithms. In: Parallel Problem Solving from Nature, PPSN IX, Berlin, Germany, pp. 920–929. Springer, Heidelberg (2006)
23. Coelho, A.L.V., de Oliveira, D.G.: Dynamically tuning the population size in particle swarm optimization. In: Proceedings of the ACM Symposium on Applied Computing (SAC 2008), pp. 1782–1787. ACM Press, New York (2008)
24. Auger, A., Hansen, N.: A restart CMA evolution strategy with increasing population size. In: Proceedings of the IEEE Congress on Evolutionary Computation (CEC 2005), pp. 1769–1776. IEEE Press, Piscataway, NJ (2005)
25. Chen, J., Qin, Z., Liu, Y., Lu, J.: Particle swarm optimization with local search. In: Proceedings of the International Conference on Neural Networks and Brain (ICNN&B 2005), pp. 481–484. IEEE Press, Piscataway, NJ (2005)
26. Gimmler, J., Stützle, T., Exner, T.E.: Hybrid particle swarm optimization: An examination of the influence of iterative improvement algorithms on performance. In: Dorigo, M., Gambardella, L.M., Birattari, M., Martinoli, A., Poli, R., Stützle, T. (eds.) ANTS 2006. LNCS, vol. 4150, pp. 436–443. Springer, Heidelberg (2006)
27. Das, S., Koduru, P., Gui, M., Cochran, M., Wareing, A., Welch, S.M., Babin, B.R.: Adding local search to particle swarm optimization. In: Proceedings of the IEEE Congress on Evolutionary Computation (CEC 2005), pp. 428–433. IEEE Press, Piscataway, NJ (2005)
28. Petalas, Y.G., Parsopoulos, K.E., Vrahatis, M.N.: Memetic particle swarm optimization. Annals of Operations Research 156(1), 99–127 (2007)

Optimised Search Heuristic Combining Valid Inequalities and Tabu Search

Susana Fernandes[1,*] and Helena R. Lourenço[2,**]

[1] Universidade do Algarve, Faro, Portugal
sfer@ualg.pt
[2] Univertitat Pompeu Fabra, Barcelona, Spain
helena.ramalhinho@upf.edu

Abstract. This paper presents an Optimised Search Heuristic that combines a tabu search method with the verification of violated valid inequalities. The solution delivered by the tabu search is partially destroyed by a randomised greedy procedure, and then the valid inequalities are used to guide the reconstruction of a complete solution. An application of the new method to the Job-Shop Scheduling problem is presented.

Keywords: Optimised Search Heuristic, Tabu Search, GRASP, Valid Inequalities, Job-shop Scheduling.

1 Introduction

Recently a new class of hybrid procedures, that combine local search based (meta) heuristics and exact algorithms of the operations research field, have been designed to find solutions for combinatorial optimisation problems. Fernandes and Lourenço [1] designated these methods by Optimised Search Heuristics (OSH). Different combinations of different procedures are present in the literature, and there are several applications of the OSH methods to different problems (see the web page of Fernandes and Lourenço (2007))[1].

We present an OSH procedure that uses valid inequalities to reconstruct a local optimal solution that has been partially destroyed. We first build a feasible solution with a GRASP procedure and perform a tabu search to get a "good" local optimum. To continue searching the solution space we perturb the current solution partially destroying it and then rebuilding it. A greedy randomised method is used to delete some elements from the local optimal solution. We then test the existence of violated valid inequalities by the partial solution. These allow us to establish a new search path for rebuilding a complete feasible solution, and hopefully lead us to an attractive unexplored region of the solution space. We named this procedure Tabu_VVI from **Tabu** with **V**iolated **V**alid **I**nequalities.

---

* The work of Susana Fernandes is supported by the the program POCI2010 of the Portuguese Fundação para a Ciência e Tecnologia.

** The work of Helena R. Lourenço is supported by Ministerio de Educacion y Ciencia, Spain, MEC-SEJ2006-12291.

[1] http://www.econ.upf.edu/~ramalhin/OSHwebpage/index.html

M.J. Blesa et al. (Eds.): HM 2008, LNCS 5296, pp. 87–101, 2008.

The idea of this new method is to mimic the cuts in integer programming, letting the violated valid inequalities cut off regions of the solution space where the objective function would have a value not better than the one of the current solution. This way the search is guided from a local optimal solution to a higher-quality region of the search space.

The procedure is illustrated with an application to the job-shop scheduling problem.

This paper is organized as follows: we start by presenting a literature review and motivation, proceed introducing the job-shop scheduling problem and continue describing the application of the Tabu_VVI to it. Computational results are presented along with comparisons to other OSH methods applied to the job shop problem and also to the state of the art tabu search algorithm of Nowicki and Smutnicki [2].

2   Literature Review and Motivation

In the literature we can find a few works combining metaheuristics with exact algorithms applied to the job-shop scheduling problem, designated as Optimised Search Heuristics (OSH) by Fernandes and Lourenço [1].

Chen, Talukdar and Sadeh [3] and Denzinger and Offermann [4] design parallel algorithms that use asynchronous agents information to build solutions; some of these agents are genetic algorithms, others are branch-and-bound algorithms.

Tamura, Hirahara, Hatono and Umano [5] design a genetic algorithm where the fitness of each individual, whose chromosomes represent each variable of the integer programming formulation, is the bound obtained solving lagrangean relaxations.

The works of Adams, Balas and Zawack [6], Applegate and Cook [7], Caseau and Laburthe [8], Balas and Vazacopoulos [9] and Pezzella and Merelli [10] all use an exact algorithm to solve a sub problem within a local search heuristic for the job-shop scheduling. Caseau and Laburthe [8] build a local search where the neighbourhood structure is defined by a subproblem that is exactly solved using constraint programming. Applegate and Cook [7] develop the shuffle heuristic. At each step of the local search the processing orders of the jobs on a small number of machines is fixed, and a branch-and-bound algorithm completes the schedule. The shifting bottleneck heuristic, due to Adams, Balas and Zawack [6], is an iterated local search with a construction heuristic that uses a branch-and-bound to solve the subproblems of one machine with release and due dates. Balas and Vazacopoulos [9] work with the shifting bottleneck heuristic and design a guided local search, over a tree search structure, that reconstructs partially destroyed solutions. The procedure of Pezzella and Merelli [10] is a tabu search that uses a branch-and-bound to solve one-machine subproblems; both at the construction of the initial solution and at a re-optimisation phase of the algorithm.

Lourenço [11] and Lourenço and Zwijnenburg [12] use branch-and-bound algorithms to strategically guide an iterated local search and a tabu search algorithm. The diversification of the search is achieved by applying a branch-and-bound method to solve a one-machine scheduling subproblem obtained from the incumbent solution.

In the work of Schaal, Fadil, Silti and Tolla [13] an interior point method generates initial solutions of the linear relaxation. A genetic algorithm finds integer solutions. A cut is generated based on the integer solutions found and the interior point method is applied again to diversify the search. This procedure is defined for the generalized job-shop problem.

The interesting work of Danna, Rothberg and Le Pape [14] "applies the spirit of metaheuristics" in an exact algorithm. Within each node of a branch-and-cut tree, the solution of the linear relaxation is used to define the neighbourhood of the current best feasible solution. The local search consists in solving the restricted MIP problem defined by the neighbourhood.

We are especially interested in combinations of exact and heuristic methods where the exact procedures can be used to strategically guide the heuristic ones. In this paper we mimic the cutting plane algorithms using the verification of the existence of violated valid inequalities to guide the search in the solution space. We are not aware of any other works using this methodology.

We chose to apply this new method to the job-shop scheduling problem because it is considered a particularly hard combinatorial optimisation problem of the NP-hard class, and so a few methods that combine exact and heuristic procedures have already been design to handle it.

3   The Job Shop Scheduling Problem

The job-shop scheduling problem (JSSP) has been known to the operations research community since the early 50's [15]. It is considered a particularly hard combinatorial optimisation problem of the NP-hard class [16] and it has numerous practical applications; which makes it an excellent test problem for the quality of new scheduling algorithms. These are main reasons for the vast literature on both exact and heuristic procedures applied to this scheduling problem.

The job-shop scheduling problem considers a set of jobs to be processed on a set of machines. Each job is defined by an ordered set of operations and each operation is assigned to a machine with a predefined constant processing time (pre-emption is not allowed). The order of the operations within the jobs and its correspondent machines are fixed a priori and independent from job to job. To solve the problem we need to find a sequence of operations on each machine respecting some constraints and optimising some objective function. It is assumed that two consecutive operations of the same job are assigned to different machines, that each machine can only process one operation at a time and that different machines cannot process the same job simultaneously. We will adopt the maximum of the completion time of all jobs – the makespan – as the objective function.

A common representation for the job-shop problem is the disjunctive graph $G = (O, A, E)$ [17]; where O is the node set, corresponding to the set of operations with two dummy operations; 0 representing the source node and $o + 1$ the sink node; A is the set of arcs between consecutive operations of the same job, and E is the set of edges between operations processed by the same machine. For every node j of $O\backslash\{0, o + 1\}$ there are unique nodes i and l such that arcs

(i, j) and (j, l) are elements of A. Node i is called the job predecessor of node j - $jp(j)$ and l is the job successor of j - $js(j)$. Finding a solution to the job shop scheduling problem means replacing every edge of E with a directed arc, constructing an acyclic directed graph $D_S = (O, A \bigcup S)$ where $S = \bigcup_k S_k$ corresponds to an acyclic union of sequences of operations for each machine k. The optimal solution is the one represented by the graph D_S having the critical path from 0 to $o + 1$ with the smallest length or makespan.

4   Tabu_VVI Applied to the JSSP

The algorithm Tabu_VVI has two main stages. The first stage consists of building a feasible solution, and executing the tabu search procedure starting from it. The second stage consists of a large step followed by the tabu search, and it is repeated for a predefined number of iterations. The large step partially destroys the solution delivered by the tabu search, looks for violated valid inequalities that enforce some order between unscheduled operations, and then rebuilds a complete solution respecting those established orders. The information about the algebraic structure of the problem within the valid inequalities is used to guide the search. The idea is to perturb the current complete solution achieving diversification and leading the search method to new unexplored regions of the solution space.

The main loop of the algorithm is stopped either when the lower bound of the instance is achieved (LB), or a predefined maximum number of iterations are executed without improving the upper bound (UB). Figure 1 shows a not detailed and simplified pseudo-code of algorithm Tabu_VVI.

4.1   Building a Feasible Solution

We first build a feasible solution using a GRASP_B&B algorithm [18]. It is a simple heuristic that includes a branch-and-bound method at the building phase of a GRASP procedure. A GRASP [19] is an iterative process where each iteration consists of two steps: a randomised building step of a greedy nature and a local search step. The branch-and-bound is used in the building step to solve subproblems of single machine scheduling problems. The neighbourhood of the local search uses the notions of blocks of critical operations, defining critical pairs of operations belonging to the same block, and performing forward and backward moves on them. A block of critical operations is a maximal ordered set of consecutive operations of a critical path (in the disjunctive graph that represents the solution), sharing the same machine. Let $L(i, j)$ denote the length of the critical path from node i to node j.

Two operations u and v form a forward critical pair (u, v) if:

a) they both belong to the same block;
b) v is the last operation of the block;
c) operation $js(v)$ also belongs to the same critical path or v is the last operation of the job;

d) the length of the critical path from v to $o+1$ is not less than the length of the critical path from $js(u)$ to $o+1$ $(L(v,o+1) \geq L(js(u),o+1))$.

Two operations u and v form a backward critical pair (u,v) if:

a) they both belong to the same block;
b) u is the first operation of the block;
c) operation $jp(u)$ also belongs to the same critical path or u is the first operation of the job;
d) the length of the critical path from 0 to u, including the processing time of u, is not less than the length of the critical path from 0 to $jp(v)$, including the processing time of $jp(v)$ $(L(0,u) + p_u \geq L(0,jp(v)) + p_{jp(v)})$.

Conditions d) are included to guarantee that all moves lead to feasible solutions [9]. A forward move is executed by moving operation u to be processed immediately after operation v. A backward move is executed by moving operation v to be processed immediately before operation u.

For a detailed description of the GRASP_B&B algorithm please refer to [18].

4.2 Tabu Search

A tabu search procedure [20,21] is a local search procedure that inspects the whole neighbourhood of a current solution x and executes the move that produces the best neighbour $ybest$. The value of $ybest$ may be worse than the one of x, so the move that goes back from $ybest$ to x becomes forbbiden, named tabu moves. The set of tabu moves is updated in every iteration of the method, so

Tabu_VVI

```
xi = GRASP_B&B(runs)
x = TabuSearch(xi)
UB =makespan(x)
xb = x
while((UB > LB) and (#iterations without improvement < max #iterations))
    xd = Destroy(x)
    xd = FindValidInequalities(xd)
    x = Rebuild(xd)
    x = TabuSearch(x)
    if(makespan(x) < UB)
        update UB
        xb = x
    endif
endwhile
return(xb)
```

Fig. 1. Outline of Tabu_VVI: (xi) - initial feasible solution, (x) - current complete solution, (xd) - partially destroyed solution, (xb) - best solution, (LB) - lower bound derived from the makespan of the first bottleneck machine

the neighbourhood definition is dynamically updated. The procedure stops after a predefined number of iterations have been performed without improving the best solution found.

In order to implement a simple tabu search procedure we need to define the neighbourhood structure, the tabu length that defines how long will a move remain tabu, and an aspiration criterion, to be able to execute moves abusively considered tabu. (this abuse happens because we do not keep track of the pair of solutions before and after a move, but only of some features of the move).

The neighbourhood structure of the tabu search implemented is the same used in the local search of the GRASP_B&B [18]. But this time we keep track of those moves rejected by conditions d) because they could produce a cycle in the disjunctive graph, thus leading to an infeasible solution. When the neighbourhood is empty, we look in these rejected moves for feasibility and execute the one that generates the best feasible solution. If none of the rejected moves produces a feasible solution we then execute the tabu move that would remain tabu for the shortest number of iterations.

The number of iterations a move (performed on solution x) stays tabu – the tabu length – is defined so it depends on the size of the neighbourhood of solution x. If a solution x has many neighbours, the reverse move of the one executed to leave from it stays tabu for a longer number of iterations than the reverse move of the one executed to leave from a solution y with a smaller neighbourhood. This way we state that the possibility of returning to a previously visited solution is not equal for every solution but depends on the number of neighbours it has.

The aspiration criterion allows a tabu move to be executed if the value of the resulting solution is better than the best one found so far.

Every time the tabu search improves the best known solution we apply an intensification scheme that consists in repeating the tabu search, this time duplicating the number of allowed iterations without improvement.

4.3 Large Step

Partially destroying a solution. The tabu search module of the algorithm provides a local optimal solution and its makespan is an upper bound for the optimal value. This solution is then perturbed using a greedy randomised method to eliminate the sequences of processing operations of some machines. Considering the acyclic directed graph that represents the solution, arcs connecting operations processed by the same machine are deleted. This method is biased toward machines that, when their sequence of processing operations is deleted, lead to a bigger reduction on the makespan of the solution. We keep "deleting" machines (destroying the sequence for processing the operations) until the makespan of the resulting partial solution is less than the upper bound.

After a predefined maximum number of global iterations are executed without improving the best solution found, the algorithm continues, for the same amount of iterations, this time choosing to "delete" machines that lead to the smallest reduction on the makespan. While the best solution found keeps being updated, we keep running the algorithm, alternating the criteria for "deleting" machines from the solution.

Finding violated valid inequalities. Having a partial solution and an upper bound (UB) for the optimal value, we then test the existence of violated valid inequalities. These allow us to establish some orders between operations of each unscheduled machine.

The procedure looks for violated valid inequalities for every machine whose sequence of operations is not present on the current partial solution. The process cycles through all the "deleted" machines and is repeated until no more orders between operations are set.

We use the same inequalities that were used in the branch-and-bound algorithms of Carlier and Pinson [22] and Applegate and Cook [7].

Let α be a machine of the instance whose sequence of processing the operations was deleted from the solution, and S_α any given sub-set of the operations processed by α. Every operation i has an earliest possible starting time - e_i, a processing time - p_i and a minimum completion time after it is processed - f_i.

If for any given set S_α and any given operation $i \in S_\alpha$, $\min_{j \in S_\alpha \setminus \{i\}} \{e_j\} + \sum_{j \in S_\alpha} p_j + \min_{j \in S_\alpha} \{f_j\} \geq UB$ then, to be possible to reduce the upper bound, operation i must be processed on α before any other operation in S_α. The inverse inequality $\min_{j \in S_\alpha} \{e_j\} + \sum_{j \in S_\alpha} p_j + \min_{j \in S_\alpha \setminus \{i\}} \{f_j\} \geq UB$ states that operation i must be processed on α after any other operation in S_α.

Let C_α be the set of operations not yet ordered for machine α, $E_\alpha \subseteq C_\alpha$ the sub-set of operations that could be scheduled first, and $F_\alpha \subseteq C_\alpha$ the subset of operations that could be scheduled last. If there is an operation $i \in E_\alpha$ such that $e_i + \sum_{j \in C_\alpha} p_j + \min_{j \in F_\alpha} \{f_j\} \geq UB$ then i can be removed from E_α. If E_α contains only one operation, then it must be processed on α before any other operation in C_α. The reverse inequality $\min_{j \in E_\alpha} \{e_j\} + \sum_{j \in C_\alpha} p_j + f_i \geq UB$ states that i cannot be scheduled after all the other operations in C_α, and should be removed from F_α.

Not all the sub-sets S_α are inspected when looking for violated valid inequalities that allow us to fix orders between operations of one machine, as it would be too computationally expensive. A reduced number of sub-sets are formed including operations by its decreasing values of starting and completion times.

If when looking for violated valid inequalities we find none, then we reintroduce a deleted machine in the solution and we look again for violated valid inequalities. The machine to add to the solution is chosen randomly. If the violated valid inequalities lead to incompatible sequences of operations, this means we cannot improve the upper bound (UB) with the set of sequenced machines, and another machine is deleted from the solution. If this happens repeatedly and the solution becomes empty, then the current complete solution is optimal.

Rebuilding a complete solution. The solution is reconstructed including the sequence of operations of one machine at a time. The order of adding the sequences in the machines to the solution is the same as for the elimination. The first machine to be re-included in the solution is the one that was first removed, and so on. The schedule of operations for each machine is determined using a modified version of the Schrage algorithm [23] that considers pre-defined orders between operations. Each time the sequence of operations of a machine is re-included in the solution, a restricted local search is executed, where it is forbidden to change orders fixed by the valid inequalities. When a new sequence of operations is included, we look for new violated valid inequalities in all remaining unscheduled machines, trying to fix more orders between operations.

After the solution is complete, local search is executed.

5   Computational Results

We have tested the algorithm Tabu_VVI on 132 benchmark instances: abz5-9 [6], ft6, ft10, ft20 [24], la01-40 [25], orb01-10 [7], swv01-20 [26], ta01-50 [27] and yn1-4 [28] [2]. The size of the instances is measure by the number of operations (equal to the number of jobs times the number of machines). The instances have different sizes: ft6 is the smaller one with 6×6 operations; la01-05 have 10×5; la06-10 have 15×5; ft20 and la11-15 have 20×5; abz5-6, ft10, la16-20 and orb01-10 have 10×10; la21-25 have 15×10; la26-30 and swv01-05 have 20×10; la36-40 and ta01-10 have 15×15; abz7-9, swv06-10 and ta11-20 have 20×15; ta31-40 and yn1-4 have 20×20; the bigger ones are ta41-50 with 30×20 operations.

An optimal solution has already been found for 83 of these instances; namely abz5-7, ft6, ft10, ft20, la01-40, orb01-10, swv01-02, swv05, swv13-14, swv16-20, ta01-10, ta14, ta17, ta31, ta35-36 and ta38-39.

We have tested a few slightly different versions of the method Tabu_VVI. Within the tabu search module, different values of the tabu length parameter were tested: equal to the number of neighbours; half of it and the double of it. Also inside the tabu search module, we have tested not to look for those moves rejected by conditions d), so when a neighbourhood is empty the eligible tabu move is always the one executed. The number of tabu iterations allowed without improving the best solution was set to the number of operations of each instance. Within the rebuild module, we have also tested to build the sequence of processing operations in one machine using a branch-and-bound method instead of just the priority rule of the Schrage algorithm. The orders between operations that were fixed by the find violated valid inequalities module are always respected.

At the first stage of the method Tabu_VVI, the GRASP_B&B algorithm was run for 10 iterations to generate the initial feasible solution and tabu search was run for 100 iterations without improvement.

The algorithm has been run on a Pentium 4 CPU 2.80 GHz and coded in C.

---

[2] These instances can be found in http://people.brunel.ac.uk/ mastjjb/jeb/orlib/files/ files jobshop1.txt and jobshop2.txt

In order to measure the performance of the algorithm we use the percentage of relative error to the lower bound - RE_{LB} (or to the optimum if the problem is closed). $f(x)$ stands for the makespan of the best solution found.

$$RE_{LB}(x) = 100\% \times \frac{f(x) - LB}{LB}$$

The next table 1 presents the performance of two variants of the algorithm that we found most successful, considering the sum of the RE_{LB} for all the instances tested, and also a column with the best results over all the 15 variants tested. The two chosen variants are tabu_mv_inf, the variant described earlier, and tabu_mv_bb, where moves rejected by conditions d) are not considered and branch-and-bound is used in the rebuilding module. The tabu tenure is set to be equal to the number of neighbours in both variants.

The table shows the average values (over a group of instances) of the RE_{LB} and the time in seconds to the best solution found.

Table 1. Results by Tabu_VVI: variants tabu_mv_inf and tabu_mv_bb, and the best of all variants, for all groups of instances, in average percentage of the relative error to the lower bound, and the average time to the best, in seconds

instances	tabu_mv_inf		tabu_mv_bb		best_all_variants	
	$avg(RE_{LB})$	$avg(time)$	$avg(RE_{LB})$	$avg(time)$	$avg(RE_{LB})$	$avg(time)$
abz	2.11	63.77	1.93	61.02	1.71	81.46
ft	0	11.72	0	0.58	0	0.15
la01-05	0	0.12	0	0.12	0	0.03
la06-10	0	0.02	0	0.03	0	0.02
la11-15	0	0.04	0	0.05	0	0.04
la16-20	0	1.79	0	1.67	0	0.44
la21-25	0.11	23.13	0.06	14.80	0	7.94
la26-30	0.29	54.12	0.26	40.88	0.17	83.39
la31-35	0	0.38	0	0.39	0	0.27
la36-40	0.47	22.68	0.22	33.50	0.05	57.08
orb	0.23	7	0.09	14.13	0	4.30
swv01-05	2.89	88.05	2.93	120.43	2.33	127.91
swv06-10	8.89	336.94	9.51	204.27	8.06	281.64
swv11-15	1.78	1734.51	2.03	825.21	1.41	1854.58
swv16-20	0	1.58	0	1.64	0	1.58
yn	7.49	339.33	7.91	73.61	7	163.95
ta01-10	0.63	77.72	0.81	67	0.24	49.52
ta11-20	3.47	54.20	3.70	86.60	3.12	177.24
ta21-30	6.51	319.27	6.60	269.97	5.96	319.02
ta31-40	1.79	230.90	1.60	258.49	1.26	220.62
ta41-50	6.04	650.87	5.88	559.71	5.47	1016.21

Table 2. Results by variants tabu_mv_inf and tabu_mv_bb of Tabu_VVI, and the algorithm of Caseau and Laburthe, in average percentage of the relative error to the lower bound, and the average time to the best, in seconds

instances	Tabu_VVI				CL	
	tabu_mv_inf		tabu_mv_bb			
	$avg(RE_{LB})$	$avg(time)$	$avg(RE_{LB})$	$avg(time)$	$avg(RE_{LB})$	$avg(time)$
abz	2.11	63.77	**1.93**	61.02	2.57	112.67
ft	0	11.72	0	0.58	0	112
la01-05	0	0.12	0	0.12	0	3.80
la06-10	0	0.02	0	0.03	0	0.75
la11-15	0	0.04	0	0.05	0	27
la16-20	0	1.79	0	1.67	0	25.08
la21-25	0.11	23.13	**0.06**	14.80	0.11	551.40
la26-30	0.29	54.12	**0.26**	40.88	0.47	4322.25
la31-35	0	0.38	0	0.39	0	2108.40
la36-40	0.47	22.68	**0.22**	33.50	0.37	2476.40
orb	0.23	7	**0.09**	14.13	1.66	111.11

We have found a new upper bound, 1765, for instance swv10 in 101 seconds.

The values of best known lower and upper bounds were gathered from the paper of Jain and Meeran [15] and the papers of Nowicki and Smutnicki [2], [29], [30].

5.1   Comparison to Other OSH Methods

The optimised search methods applied to the job-shop scheduling problem, that we know of and have mentioned in the literature review, are only applied to the older and easier instances of the problem, except for the works of Balas and Vazacopoulos [9] and Pezzella and Merelli [10], that will be treated separately.

The method of Danna, Rothberg and Le Pape [14] is applied to instances of the weighted-tardiness version of the problem, and the work of Schaal, Fadil, Silti and Tolla [13] is applied to the generalised scheduling problem.

Our method, Tabu_VVI is better for all the comparable instances (except for one or two exceptions), in quality of the solutions and in computational time, then the works of Chen [3], Denzinger and Offermann [4], Tamura, Hirahara, Hatono and Umano [5], Adams, Balas and Zawack [6], Applegate and Cook [7], Lourenço [11] and Lourenço and Zwijnenburg [12]. In table 2 we show the comparison results to the work of Caseau and Laburthe (named CL), because it is the best of these methods and also because it is the one that presents results for more instances. Their algorithm was run on a SunSparc 10 machine. The running times for their method are not scaled for our PC. Nonetheless we state our algorithm is faster and achieves better quality solutions.

Comparison to Guided Local Search. The guided local search procedure of Balas and Vazacopoulos [9] designs a search procedure based on local improvements and accepting non improving moves, using structures of neighbourhood trees. Each neighbourhood tree corresponds to a cycle of the guided local search procedure. Each node of the tree stores a solution and each edge connects neighbour solutions. Feasible solutions are built solving to optimality by branch-and-bound all one-machine subproblems (like the shifting bottleneck heuristic [6]). After a few cycles of neighbourhood trees, the procedure randomly destroys the best solution found; deleting the sequence of operations for some machines, and then reconstructs the partially destroyed solution repeating the all process.

Here we compare our best results to their best reported version SB-RGSL10, which stands for shifting bottleneck with randomised guided local search. The 10 means the number of times the all process is repeated. We call it BZ. Their algorithm was run on a SunSparc 30 machine. The comparison results between algorithms Tabu_VVI and BZ are shown in table 3. Although we used different computers and their running times are not scaled for our PC, we can still say that our method is always faster then BZ. Quality values that win the comparison are shown in bold.

Comparison to the Tabu Search with Shifting Bottleneck. The procedure of Pezzella and Merelli [10] combines tabu search with the shifting bottleneck heuristic. The later is used to build the initial solution, and also at the

Table 3. Results by the best of all variants of Tabu_VVI and the best variant of the algorithm of Balas and Vazacopoulos; in average percentage of the relative error to the lower bound, and the average time per run to the best, in seconds

instances	Tabu_VVI		BZ	
	$avg(RE_{LB})$	$avg(time)$	$avg(RE_{LB})$	$avg(time)$
la01-05	0	0.03	0	5.9
la16-20	0	0.44	0	47
la21-25	0	7.94	0	139.6
la26-30	**0.17**	83.4	0.19	121.6
la36-40	0.05	57.1	**0.03**	278
orb	**0**	4.30	0.10	80.18
swv01-05	2.33	128	**2.02**	1290
swv06-10	**8.06**	282	9.64	2917
swv11-15	**1.41**	1855	2.12	9173
yn	7	164	**5.96**	5938
ta01-10	**0.24**	49.5	0.25	1182
ta11-20	**3.12**	177	3.34	3383
ta21-30	**5.96**	319	6.57	4377
ta31-40	1.26	221	**1.13**	5069
ta41-50	**5.47**	1016	5.71	10726

re-optimisation phase of the algorithm. Whenever the tabu search cycle improves the best known solution, the procedure deletes the sequence of operations of all critical machines (machines with operations in the critical path). After shifting bottleneck rebuilds the solution, the tabu search is repeated. The tabu search module uses a dynamic management of three different neighbourhood structures and a tabu list of variable size, dependent of how many tabu iterations have been executed. The algorithm, that we name PM, was run on a Pentium 133 MHz. Table 4 shows the comparison results between algorithms Tabu_VVI and PM. Quality values that win the comparison are shown in bold.

Table 4. Results by the best of all variants of Tabu_VVI and the algorithm of Pezzella and Merelli; in average percentage of the relative error to the lower bound, and the average time to the best, in seconds

instances	Tabu_VVI		PM	
	$avg(RE_{LB})$	$avg(time)$	$avg(RE_{LB})$	$avg(time)$
abz	**1.71**	81.5	2.23	151
ft	0	0.15	0	65
la01-05	0	0.03	0	9.8
la06-10	0	0.02	0	-
la11-15	0	0.04	0	-
la16-20	0	0.44	0	61.5
la21-25	**0**	7.94	0.1	115
la26-30	**0.17**	83.4	0.46	105
la31-35	0	0.27	0	-
la36-40	**0.05**	57.1	0.58	141
ta01-10	**0.24**	49.5	0.45	2175
ta11-20	**3.12**	177	3.47	2526
ta21-30	**5.96**	319	6.52	34910
ta31-40	**1.26**	221	1.92	141333
ta41-50	**5.47**	1016	6.04	11512

5.2   Comparison to State of the Art Procedure - Tabu Search with Path-Relinking

Along with the guided local search procedure of Balas and Vazacopoulos [9], and the tabu search with shifting bottleneck of Pezzella and Mirelli [10], one other procedure, due to Nowicki and Smutnicki [2], forms the group of three procedures that are the best up to date methods applied to the job-shop scheduling problem.

The procedure of Nowicki and Smutnicki performs path-relinking between elite solutions found by a tabu search module. The solutions achieved by the path-relinking are then used as starting points for new cycles of the tabu search; the set of elite solutions is updated and the all process is repeated. We can say that the path-relinking works as the diversification strategy of the tabu search.

The algorithm uses a data structure specially designed for the application of this method to the job-shop scheduling problem. The instances of Taillard [27] were used to study the distribution of the local optima solutions in the solution space; and this study supported the design of this method. The algorithm, that we name NS, was run on a Pentium 900 MHz. Unlike all other procedures, the computational times reported by the authors do not include the time needed to build the initial solutions. Table 5 shows the comparison results between algorithms Tabu_VVI and NS. After running for approximately the same amount of time, Tabu_VVI achieves solutions with quality very close to the results of NS.

Table 5. Results by the best of all variants of Tabu_VVI and the algorithm of Nowicki and Smutnicki; in average percentage of the relative error to the lower bound, and the average time to the best, in seconds

instances	Tabu_VVI		NS	
	$avg(RE_{LB})$	$avg(time)$	$avg(RE_{LB})$	$avg(time)$
swv01-05	2.33	128	1.01	462
swv06-10	8.06	282	7.49	514
swv11-15	1.41	1855	0.51	360
yn	7	164	5.18	510
ta01-10	0.24	50	0.11	26
ta11-20	3.12	177	2.81	108
ta21-30	5.96	319	5.68	328
ta31-40	1.26	221	0.78	341
ta41-50	5.47	1016	4.7	975

6 Conclusions

We have developed a powerful, fast and innovative optimised search heuristic to solve combinatorial optimisation problems. It uses an exact technique from the operations research field to guide the search process of a metaheuristic. The procedure, named Tabu_VVI, uses the verification of violated valid inequalities as a diversification strategy of a tabu search procedure. The idea of this new method is to mimic the cuts in integer programming, letting the violated valid inequalities discard the current solution and guide the search from a local optimal solution to a more quality region of the search space.

The procedure was illustrated with an application to the job-shop scheduling problem. We presented some computational results for a large set of benchmark instances, along with comparisons to other similar and successful works. Our new method, Tabu_VVI, always performs better than other methods that combine exact and heuristic procedures. It compares most favourably to two other leading methods for solving the job-shop scheduling problem; the guided local search of Balas and Vazacopoulos [9] and the tabu search with shifting bottleneck of Pezzella and Mirelli [10]. When compared to the state of the art tabu search of

Nowicki and Smutnicki [2], after running for approximately the same amount of time, Tabu_VVI achieves solutions with quality very close to theirs.

References

1. Fernandes, S., Lourenço, H.R.: Optimized search heuristics. Technical report, Universitat Pompeu Fabra (2007)
2. Nowicki, E., Smutnicki, C.: An advanced tabu search algorithm for the job shop problem. Journal of Scheduling 8, 145–159 (2005)
3. Chen, S., Talukdar, S., Sadeh, N.: Job-shop-scheduling by a team of asynchronous agent. In: IJCAI 1993 Workshop on Knowledge-Based Production, Scheduling and Control, Chambéry, France (1993)
4. Denzinger, J., Offermann, T.: On cooperation between evolutionary algorithms and other search paradigms. In: 1999 Congress on Evolutionary Computation (CEC). IEEE Press, Los Alamitos (1999)
5. Tamura, H., Hirahara, A., Hatono, I., Umano, M.: An approximate solution method for combinatorial optimisation. Transactions of the Society of Instrument and Control Engineers 130, 329–336 (1994)
6. Adams, J., Balas, E., Zawack, D.: The shifting bottleneck procedure for job shop scheduling. Management Science 34(3), 391–401 (1988)
7. Applegate, D., Cook, W.: A computational study of the job-shop scheduling problem. ORSA Journal on Computing 3(2), 149–156 (1991)
8. Caseau, Y., Laburthe, F.: Disjunctive scheduling with task intervals. Technical Report LIENS 95-25, Ecole Normale Superieure Paris (July 1995)
9. Balas, E., Vazacopoulos, A.: Guided local search with shifting bottleneck for job shop scheduling. Management Science 44(2), 262–275 (1998)
10. Pezzella, F., Merelli, E.: A tabu search method guided by shifting bottleneck for the job shop scheduling problem. European Journal of Operational Research 120, 297–310 (2000)
11. Lourenço, H.R.: Job-shop scheduling: Computational study of local search and large-step optimization methods. European Journal of Operational Research 83, 347–367 (1995)
12. Lourenço, H.R., Zwijnenburg, M.: Combining large-step optimization with tabu-search: Application to the job-shop scheduling problem. In: Osman, I.H., Kelly, J.P. (eds.) Meta-heuristics: Theory & Applications. Kluwer Academic Publishers, Dordrecht (1996)
13. Schaal, A., Fadil, A., Silti, H.M., Tolla, P.: Meta heuristics diversification of generalized job shop scheduling based upon mathematical programming techniques. In: CP-AI-OR 1999 (1999)
14. Danna, E., Rothberg, E., Pape, C.L.: Exploring relaxation induced neighborhoods to improve MIP solutions. Mathematical Programming, Ser. A 102, 71–90 (2005)
15. Jain, A.S., Meeran, S.: Deterministic job shop scheduling: Past, present and future. European Journal of Operational Research 133, 390–434 (1999)
16. Garey, M.R., Johnson, D.S.: Computers and Intractability: A Guide to the Theory of NP-Completeness. Freeman, San Francisco (1979)
17. Roy, B., Sussman, B.: Les problems d'ordonnancement avec constraintes disjonctives. Technical report, Notes DS 9 bis, SEMA, Paris (1964)
18. Fernandes, S., Lourenço, H.R.: A GRASP and branch-and-bound metaheuristic for the job-shop scheduling. In: Cotta, C., van Hemert, J. (eds.) EvoCOP 2007. LNCS, vol. 4446, pp. 60–71. Springer, Heidelberg (2007)

19. Feo, T., Resende, M.: Greedy randomized adaptive search procedures. Journal of Global Optimization 6, 109–133 (1995)
20. Glover, F.: Tabu search - part i. ORSA Journal on Computing 1(3), 190–206 (1989)
21. Glover, F.: Tabu search - part ii. ORSA Journal on Computing 2(1), 4–32 (1990)
22. Carlier, J., Pinson, E.: An algorithm for solving the job-shop problem. Management Science 35(2), 164–176 (1989)
23. Schrage, L.: Solving resource-constrained network problems by implicit enumeration: Non pre-emptive case. Operations Research 18, 263–278 (1970)
24. Fisher, H., Thompson, G.L.: Probabilistic learning combinations of local job-shop scheduling rules. In: Muth, J.F., Thompson, G.L. (eds.) Industrial Scheduling, pp. 225–251. Prentice-Hall, Englewood Cliffs (1963)
25. Lawrence, S.: Resource constrained project scheduling: an experimental investigation of heuristic scheduling techniques. Technical report, Graduate School of Industrial Administration, Carnegie-Mellon University (1984)
26. Storer, R.H., Wu, S.D., Vaccari, R.: New search spaces for sequencing problems with application to job shop scheduling. Management Science 38(10), 1495–1509 (1992)
27. Taillard, E.D.: Benchmarks for basic scheduling problems. European Journal of Operational Research 64(2), 278–285 (1993)
28. Yamada, T., Nakano, R.: A genetic algorithm applicable to large-scale job-shop problems. In: Manner, R., Manderick, B. (eds.) Parallel Problem Solving from Nature 2, pp. 281–290. Elsevier Science, Brussels Belgium (1992)
29. Nowicki, E., Smutnicki, C.: Some new tools to solve the job shop problem. Technical Report 60/2002, Institute of Engineering Cybernetics, Wroclaw University of Technology (2002)
30. Nowicki, E., Smutniki, C.: A fast taboo search algorithm for the job shop problem. Management Science 42(6), 797–813 (1996)

Iterated Greedy Algorithms for a Real-World Cyclic Train Scheduling Problem

Zhi Yuan[1], Armin Fügenschuh[2], Henning Homfeld[2], Prasanna Balaprakash[1], Thomas Stützle[1], and Michael Schoch[3]

[1] IRIDIA-CoDE, Université Libre de Bruxelles (ULB), Brussels, Belgium
{zyuan,pbalapra,stuetzle}@ulb.ac.be
[2] Arbeitsgruppe Optimierung, Fachbereich Mathematik, Technische Universität Darmstadt, Darmstadt, Germany
{fuegenschuh,homfeld}@mathematik.tu-darmstadt.de
[3] Deutsche Bahn AG, Frankfurt am Main, Germany
michael.schoch@bahn.de

Abstract. In this paper, we develop heuristic algorithms for a complex locomotive scheduling problem in freight transport that arises at Deutsche Bahn AG. While for small instances an approach based on an ILP formulation and its solution by a commercial ILP solver was rather successful, it was found that effective heuristic algorithms are needed for providing better initial upper bounds and for tackling large instances. The main contribution of this paper is the development of heuristic algorithms that strongly improve over the performance of the greedy algorithm used in the previous research efforts. The development process was done on a step-by-step basis ranging from improvements over the initial greedy construction heuristic, the development of a simple local search algorithm, the further extension to an iterated greedy procedure to the adoption of population-based stochastic local search methods. Our computational results show that the iterated greedy algorithm combined with a simple local search is a powerful algorithm for this real-world freight train scheduling problem.

1 Introduction

The problem we are tackling in this paper arises in the strategic planning of the Deutsche Bahn AG (DB), the largest railway company in Germany. In particular, the problem arises in the context of a complex simulation tool that is used at DB to provide long-term simulations and future predictions of the load of the railway network. The tool can be seen as a chain of modules, where information between the modules is exchanged through data files.

Our particular problem arises in a module called *train scheduler*, which is responsible for the buildup of trains from cars. A train starts as soon as it is built, that is, when enough cars are assembled. Hence, this also means that the starting times of the trains do not follow a specific timetable; rather they follow the estimated customers demand or production. Since the locomotives

M.J. Blesa et al. (Eds.): HM 2008, LNCS 5296, pp. 102–116, 2008.

are among the most expensive resources of the operation of railroad companies, their efficient scheduling is of high importance.

The locomotive scheduling, with which we deal here, can be characterized as a vehicle routing problem with time windows, a heterogeneous fleet of vehicles (due to different types of locomotives), and cyclic departures of trains. It also includes two important additional aspects: network-load dependent travel times and the transfer of cars between trains. In earlier work [1], we have developed an integer linear programming (ILP) formulation of the problem and proposed a solution approach for the problem using a commercial ILP solver (ILOG Cplex 10 [2]). In the corresponding experimental campaign [1], we noticed that (i) the minimum number of missed car transfers is relatively easy to find, (ii) with fixed starting times, relatively large instances with up to about 1 500 trips could be solved to optimality, (iii) the models that allowed the flexible choice of starting times within some predefined time windows made the problem much more difficult to solve: the size of the instances that could successfully be tackled by the commercial ILP solver (after some additional improvements such as providing good initial feasible solutions and problem-specific cutting planes) was limited to medium sized instances with a few hundreds of trips [1]. Despite the added computational difficulty, the hardest model with flexible starting times has highly desirable properties: allowing to vary the starting times within small time windows results in a considerable reduction of the required number of locomotives and, hence, a strong reduction in the total costs [1].

We tackle the most difficult problem variant studied in [1], namely the one that uses time windows and network-load dependent travel times. For this variant, which is also the most realistic and interesting one, heuristic algorithms are required to generate good quality solutions to large instance sizes, but also ILP approaches can benefit from improved initial upper bounds for medium sized instances. In this paper, we therefore report on our research for improving upon the performance of the greedy construction heuristics that have been used in [1]. Our development process departs from these greedy heuristics, extending them step-by-step. The first extensions comprise a direct modification of the greedy heuristic by changing the way solutions are constructed. Next, we extend the construction heuristic to an iterated greedy (IG) algorithm [3] and further hybridize it with a simple iterative improvement algorithm. Our experimental results show that for computation times ranging from a few seconds to a few minutes on current CPUs, very strong improvements over the initial greedy heuristic can be obtained. A further hybridization of the IG algorithm with ant colony optimization (ACO) algorithms, however, gave rather mixed results and no further significant improvement in performance.

2   The Freight Train Scheduling Problem

It is convenient to first describe the nomenclature used in the context where this freight train scheduling problem arises.

2.1 Problem Setting

Cars. A *car* is the smallest unit to be moved; cars have to be moved from a source to a destination within the railway network. A train is composed of a set of cars. Large customers require the transport of large amounts of goods so that they order whole trains; in such a case, the route of the cars is the same as the route of the train. Individual or small sets of cars are used by smaller customers. Several such cars are then assembled into a train, moved to some intermediate destination, which is called *shunting yard*, where trains are disassembled and reassembled into new trains. We assume that that the place and timing of these transfers is known for individual cars. Note that the scheduling of locomotives needs to take care that these transfers remain feasible.

Trains. A *freight train* (also called *active trip*) consists of several *cars*. Each train has a *start* and a *destination*, which are goods stations or railroad shunting yards, and *starting times* and *arrival times*. Times can be either fixed times or be taken from some interval. In our case, trips start cyclically every 24 hours. The *trip duration* is the difference between start and arrival times. Trains vary in lengths and weight and, depending on these two characteristics, different needs on the driving power of locomotives arise. Typically, a single locomotive is enough and only rarely two pulling locomotives are required. The handling of a train involves attaching a locomotive to a train at the start and the decoupling of the locomotive at the destination. For both *coupling* processes, some time (up to 30 minutes, depending on the trip) is required for technical checks or refueling.

Locomotives. The around 30 different models of locomotives that are used at DB share many similarities and so they can be classified into a small number of different *classes*. Differences between the classes concern mainly the driving power and the motor type, diesel or electrical. Electrical locomotives can only be used on electrical tracks, whereas diesel locomotives, in principle, can drive everywhere. However, diesel soots the electrical wires, so one wants to avoid their deployment on such tracks. Hence, it is only possible to assign such locomotives to trains that have a sufficient power and the right motor type for the track.

Deadheads. A locomotive is either active, that is, pulling a train, or deadheading, that is, driving alone, without pulling a train, from the destination station of one train to the start of another train. The distance between these points and the class of locomotive determine the duration of a deadhead trip.

2.2 Formulation of the Problem

Our locomotive scheduling problem can be formulated as a cyclic Vehicle Scheduling Problem (CVSP), more specifically, as a CVSP with hard time windows and a heterogeneous fleet of vehicles (locomotives); we use the abbreviation CVSPTW in what follows. The locomotives differ in starting cost and capability, that is, the subset of customers that can be potentially served by the locomotives. Instead of starting and terminating at a depot as in standard vehicle routing problems, the locomotives are scheduled in a cyclic fashion every 24 hours. Given is also a set of customers, more specifically in our case a set of trips, that expect exactly one of the locomotives to pull the train obeying restrictions on the starting and arrival times (or time windows).

The problem can be described as follows. (A mathematical formulation of the constraints and objective as a mixed-integer programming model can be found in [1].) We denote by $\mathcal{V}$ the set of active trips and by $\mathcal{B}$ the set of locomotive types. With each trip is associated a list of locomotive types which can serve it. $\mathcal{A} = \mathcal{V} \times \mathcal{V}$ denotes the set of potential connections of pairs of trips. Note that trips are *cyclic*, which for our problem means that the trips occur every day.

The time intervals for the starting and arrival times of trips are input data. In practice, the train scheduler of the tool chain defines fixed starting and arrival times. In our model, we let the starting time t_i of trip i vary in a time window $\underline{t}_i \leq t_i \leq \bar{t}_i$. The times are defined as the minutes passed since some time zero and, hence, one has that $0 \leq t_i \leq 1439$ (a day has 1440 minutes). Note that a further constraint imposes that a trip has to start on the first day; hence, in case $\bar{t}_i$ is larger than 1439, the time interval $\underline{t}_i \leq t_i \leq \bar{t}_i$ is replaced by two time intervals $([\underline{t}_i, 1439] \cup [0, \bar{t}_i]) \cap \mathbb{Z}$.

Moreover, trains $i, j \in \mathcal{V}$ that require a car transfer from one to the other need to be synchronized. Denote $\mathcal{P}$ the set of pairs (i, j) where cars transfer from i to j. We assume $\mathcal{P}$ is valid and given by the earlier computation module. For all $(i, j) \in \mathcal{P}$ the transferred car must be picked up by j within 12 hours after the arrival of i at the shunting yard.

Our model includes network-load dependent travel times. This is important since typically at day time the average traveling speed of a freight train is lower than during nighttime; the main reason is that passenger trains have a higher priority than most freight trains, leading to frequent waiting times for the latter. To model this aspect, we partition a day into a number of time slices $\mathcal{H} = \{1, \ldots, H\}$, that is, $[0, 1439] = \bigcup_{h \in \mathcal{H}} [\underline{\psi}_h, \bar{\psi}_h]$. The starting time of a train then falls into one of the slices and the travel times are considered accordingly.

The total trip and deadhead durations $\delta_{b,(i,j)}$ for a locomotive of type b that serves both trips i and j are computed as

$$\delta_{b,(i,j)} := \delta_i^{trp} + \delta_i^{uncpl} + \delta_{b,(i,j)}^{dhd} + \delta_j^{cpl}, \tag{1}$$

where δ_i^{trp} denotes the trip duration, that is, the time the locomotive is active while pulling train i; δ_i^{uncpl} denotes the time for uncoupling the locomotive from the train at the arrival; $\delta_{b,(i,j)}^{dhd}$ denotes the time for deadheading from the end of i to the start of train j; δ_j^{cpl} denotes the time for coupling the locomotive to the train at the start of j.

Note that the driving time δ_i^{trp} is assumed to be independent of the actual locomotive class, whereas the deadhead time $\delta_{b,(i,j)}^{dhd}$ is class dependent (since diesel and electrical might use different routes). In the case of network-load dependent travel times, δ_i^{trp} gets replaced by $\delta_{i,h}^{trp}$, which gives the trip duration of train i when starting in slice h. Finally, in the case of car transfers between trains, the time for shunting a car from i to j, $\delta_{i,j}^{shnt}$, needs to be taken into account for the computation of the trip and deadhead durations.

The goal of the problem is to compute feasible starting and arrival times of the trains such that the operational costs are minimized. More concretely, two cost components are considered. Let γ_b^{cls} be the cost for a locomotive of class b, $\gamma_{b,(i,j)}^{dhd}$ be the cost of a deadhead trip from i to j for locomotive class b. Typically, the following ordering holds between the costs: $\gamma_b^{cls} \gg \gamma_{b,(i,j)}^{dhd}$, that is, the most important objective is to reduce the number of used locomotives, and then at a subsidiary level, to minimize the total distance of all the deadhead trips.

Finally, recall that in this paper we tackle the most difficult variant of the train scheduling problem from [1], that is, the variant with time windows (instead of fixed starting and arrival times) and with network-load dependent travel times. For the approach described in this paper, we do not consider the preliminary step of the minimization of the number of missed car transfers as in [1], but we fix the set of feasible car transfers as determined through some preliminary computation. The reason is that the minimum number of missed car transfers can, in practice, easily be done by an ILP approach or by a greedy heuristic [1]. Hence, we focus on the minimization of the operational costs, that is, the costs for the locomotive usage and the deadhead trips.

2.3   Benchmark Instances

For our computational tests we used eight instances, five of which were also used as test instances in [1]. The names and the characteristics of these five instances are as follows.

- A, with 42 trains, 3 locomotive classes
- B, with 82 trains, 3 locomotive classes
- C, with 120 trains, 4 locomotive classes
- KV, with 340 trains, 6 locomotive classes
- EW, with 727 trains, 6 locomotive classes

From each of these instances, we obtain additional ones by imposing different ranges for the time windows. In particular, for each of the fixed starting and arrival times we allow symmetric time windows centered at these values from the set $\{\pm 0, \pm 10, \pm 30, \pm 60, \pm 120\}$ minutes.

An additional three instances have been used for tuning the parameters of the heuristic algorithms. The characteristics of these instances are as follows.

- R1-101-3, with 101 trains, 3 locomotive classes
- R2-137-6, with 137 trains, 6 locomotive classes
- R3-295-5, with 295 trains, 5 locomotive classes

Since we used iterated F-race [4], an automated tuning tool, it was desirable to derive a larger set of instances from these three. Hence, we selected the time windows as follows. For R3-295-5, R2-137-6, and R1-101-3, we selected as time windows the integers from $[0, 150]$ that can be divided by 2, 3, and 6, respectively; this results in 75, 50, and 25 instances, respectively. Note that the larger the base instance the more derived instances are available. This was done in order to bias the choice of the parameter settings towards better solving large instances.

3   Greedy Algorithms

For effectively tackling the CVSPTW variant with network-load dependent travel times, an essential ingredient for the ILP-based approach [1] has been a randomized PGreedy-type algorithm [5]. This algorithm, called g-CVSPTW, forms the basis for the further developments and it is described next.

3.1   The g-CVSPTW Heuristic

g-CVSPTW assumes a fixed set of car transfers as input and it seeks to minimize the operating costs. g-CVSPTW iteratively generates cyclic schedules for locomotives repeating the following series of steps until all trips are scheduled.

1. **Step 1: Locomotive selection.** First, a locomotive class is chosen. The choice is based on two factors: (i) the capability N_b, which is the number of trains that can be served by a locomotive from class b, and γ_b^{cls}. Each locomotive class is scored by the ratio N_b/γ_b^{cls}. A locomotive class with the highest ratio is chosen.
2. **Step 2: Start trip selection.** Among the still unscheduled trips, we select one with the highest number of deadhead trips that use a locomotive of the same class b as selected in step 1; ties are broken randomly.
3. **Step 3: Starting time selection.** Next, the starting time t_i of the trip is taken from the interval $[\underline{t}_i, \bar{t}_i]$. t_i is chosen such that the train arrival time is the earliest possible. (Recall that since we have netload dependent driving times, we need to subdivide the interval $[\underline{t}_i, \bar{t}_i]$ in dependence of the time slices.) After fixing the starting time, the impact of this decision on other starting time windows needs to be propagated. This update is necessary, if there is some trip j with (i, j) or $(j, i) \in \mathcal{P}$. For this task, a constraint propagation algorithm was implemented to propagate the effect of the decisions through the network.
4. **Step 4: Deadhead trip selection.** Next, a shortest deadhead trip to the start of the next train is chosen for the locomotive.

g-CVSPTW starts with steps 1 and 2 and then cycles through steps 3 and 4 until no trip can be added anymore to the locomotive's schedule. In particular, the construction is stopped if no trip can be accommodated within the 24 hours cycle (recall that g-CVSPTW uses a maximum 24 hour cycle length for each locomotive).

Instead of making deterministic choices in steps 1, 2, and 4, g-CVSPTW actually uses randomized greedy values, in spirit similar to the noising method [6]. Before doing the choice, the concerned greedy values are multiplied by a value that is generated randomly according to a uniform distribution in $[10000 * (1 - NOISE), 10000]$, where $NOISE$ is a parameter from the range $[0, 1]$. Decisions are then taken deterministically based on the perturbed greedy values, breaking remaining ties randomly. The main rationale for using the randomization in the greedy heuristic is to allow the generation of a large set of different solutions.

3.2   Modified Greedy Heuristic

As a first step in the road to improved heuristic algorithms, we modified g-CVSPTW, based on the previous experience with it, resulting in the mg-CVSPTW heuristic.

The *first modification* is to add a parameter v to weigh the importance of the locomotive capability and we now use the ratio $(N_b)^v/\gamma_b^{cls}$ for the computation of the greedy values.

The *second modification* derives an additional greedy value for biasing the solution construction. It is based on the observation that the choice of a trip restricts in very different ways the number of possible trips that feasibly could be added to a locomotive cycle. In the extreme case, locomotives contain only one trip, since it is infeasible to assign more trips within the 24 hour cycle of a locomotive. (We call such trips *isolated.*) Hence, we define for each trip i its *sociability score* s_i, which counts the number of other trips that could be served by a locomotive in a same cycle with trip i; for isolated trips, this score is zero.

A *third modification* is to change the order of steps 1 and 2 in g-CVSPTW, together with a more deterministic solution construction. For this modification, now trips are pre-ordered and chosen deterministically in the given order. The ordering is defined in a lexicographic way by (i) considering the number of locomotive classes able to serve the trip (the smaller, the higher the rank) and (ii) the sociability score of each trip (the smaller the score the higher the rank). In the construction process, mg-CVSPTW first chooses the starting trip in nonincreasing order of the ranks and then fixes the locomotive class. There are two reasons why this modification was deemed to be helpful. Firstly, in this way the most constraining decisions are done first, postponing the ones with more flexibility; secondly, by selecting among a smaller set of candidates, we expected to speed-up the construction process. In fact, we observed that the computation times for the solutions construction decreased by a factor of about five.

Once the first trip and the locomotive class selection is done, the modified greedy heuristic continues iteratively with steps 3 and 4 as g-CVSPTW.

3.3   Locomotive Type Exchange Heuristic

A common way of improving solutions returned by a construction method is to improve them by some type of local search. Since the development of very effective local search algorithms can be rather time consuming, we limited the local search to locomotive class exchanges: the idea is to try to replace the locomotive of each tour by a cheaper one. The local search procedure works as follows. First, all locomotive classes are sorted according to non-decreasing costs into a list $c = \langle c_0, c_1, c_{|\mathcal{B}|} \rangle$. Then, for all cycles that use a locomotive of class i we check whether the locomotive of cost c_i can be replaced by a cheaper locomotive, that is, by one of costs $c_0, \ldots, c_{i-1}$ (starting with index 0). If this is possible, the substitution is applied and the next cyclic tour is considered.

procedure *Iterated Greedy*
   s = GenerateInitialSolution
   repeat
      s_p = DestructionPhase(s)
      s' = ConstructionPhase(s_p)
      s'' = LocalSearch(s')   % Optional local search phase
      s = AcceptanceCriterion(s, s'')
   until termination condition met

Fig. 1. Algorithmic scheme of iterated greedy with optional local search phase

4   Iterated Greedy Algorithms

As said previously, a main motivation for the randomization of the greedy steps in g–CVSPTW and in mg–CVSPTW is to allow to construct many different solutions. However, this results in independent applications of a heuristic, an approach which for many problems does not lead to excellent performance. Hence, we use *Iterated Greedy* (IG) as another way for iterating over construction heuristics. The central idea of IG is to avoid to repeat a greedy construction from scratch, but rather to keep parts of solutions between successive solution constructions [7,3]. This is done by first destructing a part of a current complete solutions and then to reconstruct from the resulting partial solution a new complete solution. This new solution is accepted as the new incumbent solution according to some acceptance criterion. For a generic algorithmic outline of IG see Figure 1.

The IG algorithm adds additional parameters. Directly linked to the IG heuristic is the choice by how much a complete candidate solution should be destructed; here, we use the parameter *destruction ratio* d. It determines the number of locomotive cycles that are removed from the current solution: if N_l is the number of locomotives in the current solution, then $\lfloor d \cdot N_l \rfloor$ randomly selected locomotives are removed from the current solution. For the re-construction of a complete solution (and the construction of the initial solution), we use the mg–CVSPTW heuristic, which proved to be more effective than g–CVSPTW.

The acceptance criterion uses the well-known Metropolis rule known from simulated annealing. A candidate solution that is better or of equal quality to the current candidate solution is deterministically accepted; a worse solution is accepted with a probability given by $\exp\{-\Delta/T\}$, where Δ is the cost difference between the new and the current solution and $T = \nu \cdot f(s^*)$, where s^* is the best solution found so far. Note that in the acceptance criterion no annealing is used.

The resulting algorithm we call IG–CVSPTW. Our algorithm uses one additional feature, which was previously not used in IG algorithms. Instead of only reconstructing one complete solution, IG–CVSPTW generates from one partial solution $I_r \geq 1$ new solutions. This is done since generating the correct time windows for the partial solutions is relatively computation time intensive due to the constraint propagation for narrowing the time windows.

Clearly, as for g–CVSPTW, it is also desirable for IG–CVSPTW to consider the hybridization of the algorithm with a local search, and, hence, we also study

Table 1. Parameter settings obtained from iterative F-race for greedy and IG heuristics with and without local search (LS). See the text for more details.

algorithm	parameter	range	selected value	
			without LS	with LS
mg-CVSPTW	$NOISE$	$[0,1]$	0.67	0.15
	v	$[0,10]$	1.8	6.3
IG	$NOISE$	$[0,1]$	0.59	0.57
	v	$[0,10]$	9.4	2.9
	d	$[0.01,1]$	0.12	0.074
	ν	$[0,0.05]$	0.046	0.019
	I_r	$[1,50]$	9	25

this feature. Adding local search to IG is straightforward by improving each reconstructed solution, as indicated by the optional LocalSearch(s') procedure in Figure 1. As for mg-CVSPTW, we use the locomotive class exchange local search.

5   Experimental Results for Greedy and Iterated Greedy

In this section, we report on the experimental results and the range of improvements that we obtained with the extensions of the greedy heuristic. Since the clearest differences in performance have been observed on the two largest base instances, we focus on these two in the discussion of the experimental results.

5.1   Experimental Setup

All the codes were written in Java and share the same data structures. The code was compiled and executed using JDK 1.6.0_05. The experiments were run on computing nodes each with two quad-core XEON E5410 CPUs running at 2.33 GHz and 8 GB RAM. Due to the sequential implementation of the algorithms, each execution makes only use of one single core.

Each algorithm was first tuned by an automatic tuning algorithm called iterative F-race [4]; the only exception is g-CVSPTW, for which the pre-fixed setting of $NOISE$ was used. The three instances R1-101-3, R2-137-6, and R3-295-5 were used as training instances for tuning, using the time windows as explained in Section 2.3. Before tuning, the order of the instances is randomized. The computation time for all tuning instances was set to 300 CPU seconds, which corresponds approximately to the time g-CVSPTW requires to generate 10 000 solutions for base instance R3-295-5. Iterative F-race is then run for a maximum of five iterations and in each iteration 100 candidate configurations are generated, that is, a total of 500 configurations for each algorithm are tested. Table 1 gives the range of values considered and the finally chosen ones by iterative F-race.

5.2   Experimental Results

For an experimental evaluation of the five algorithms, we have run each of them 30 times on the 25 test instances (for each of the five base instances 5 settings of

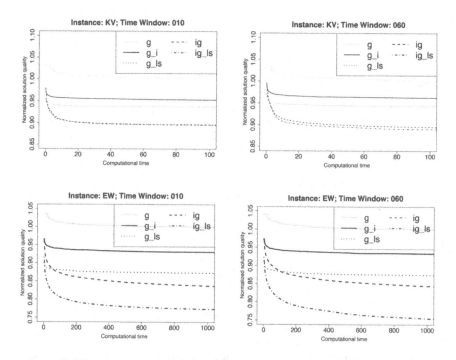

Fig. 2. Development of the solution quality over time for g-CVSPTW (g), mg-CVSPTW (g_i), mg-LS-CVSPTW (g_ls), IG-CVSPTW (ig), and IG-LS-CVSPTW (ig_ls) for instances KV-10 (top left), KV-60 (top right), EW-10 (bottom left), and EW-60 (bottom right); the numbers after the instance identifier indicate the time window range chosen. The variability of the averages is low as shown for the solutions after 100 CPU seconds (instances KV-*) and 1000 CPU seconds (instances EW-*) by the boxplots in Figure 3.

time windows have been considered). Each algorithm was run for the same maximum computation time as required by g-CVSPTW to construct 10 000 solutions. In Figure 2, we show for each of the algorithms the development of the average solution quality across 30 independent trials over time. The plots show that initially the solution quality improves very quickly and then the curves flatten off. However, especially the IG variants still show further improvements over time. The differences of the average solution quality reached by the algorithms are rather strong with the best performing one being IG-LS-CVSPTW. The boxplots, which are given in Figure 3, also indicate that the variability of each algorithm's final performance is very low. Hence, the significance of the differences could also be confirmed by statistical tests: all pairwise differences between the final solution quality reached by the algorithms are statistically significant according to the Wilcoxon test using Holms correction for the multiple comparisons; the only exception is for instance KV-10, where the differences between IG-CVSPTW and IG-LS-CVSPTW were not significant. In fact, also on few smaller instances, the differences between IG-CVSPTW and IG-LS-CVSPTW are not significant.

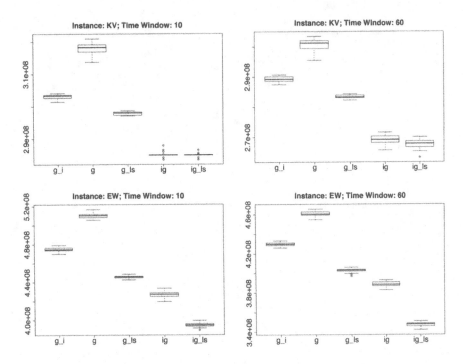

Fig. 3. Boxplots of the final performance after 100 CPU seconds for instances KV-10 (top left) and KV-60 (top right), and 1000 CPU seconds for instances EW-10 (bottom left) and EW-60 (bottom right). The y-axis shows the solution cost reached. For the explanation of the abbreviations see the caption of Figure 2.

An important result of this comparison is that the improvement of the computational results over the repeated application of g-CVSPTW is very strong. For most instances, the final costs found by IG-LS-CVSPTW were about 10% to 20% lower than those reached by g-CVSPTW. Typically, the same final solution quality as that of g-CVSPTW, was reached by IG-CVSPTW or IG-LS-CVSPTW in less than a hundredth of the maximum time taken by g-CVSPTW.

6   Iterated Ants

The strong improvements by the IG algorithms motivated us to consider extensions that might yield further performance improvements. Given the strong importance of the constructive part, we decided to consider a hybrid between two constructive SLS methods: IG and ant colony optimization (ACO) algorithms; the resulting hybrid algorithm is called *iterated ants* [8]. The central idea of iterated ants is to make each ant in the ACO algorithm follow the steps of the IG algorithm. The construction phase of the IG algorithm is then biased, as usual in ACO algorithms, by pheromones and heuristic information. The ACO approach we followed was based on a version of $\mathcal{MAX}$–$\mathcal{MIN}$ Ant System ($\mathcal{MMAS}$) that uses the pseudo-random proportional action choice rule known from Ant Colony System (see [9] for more details) and each ant follows the steps of the modified greedy construction heuristic. The pheromone trails have been associated to the choice of the next trip to be added to a locomotive's schedule; that is, a pheromone trail τ_{ij} refers to the desirability of serving trip j after having done trip i; the heuristic information is the inverse of the length of the deadhead trip from i to j. This choice matches the usual ACO approach to the TSP.

Without going into further details on the parameters involved (which were the usual ones arising in the ACO context plus the ones relevant for the IG part), we summarize our main observations. The first is deceptive, in the sense that the final configuration returned from iterated F-race had the importance of the pheromone trails set to zero (Note that the action choice rules in ACO algorithms are positively biased to choose components j for which the term $\tau_{ij}^\alpha \cdot \eta_{ij}^\beta$ is high, where α is a parameter that weighs the influence of the pheromone trails. Setting $\alpha = 0$ actually has the effect of not considering the pheromone trails at all in the solution construction.) However, the parameter weighting the influence of the heuristic information was set to its maximum value in the considered range. This is true for the variants with and without local search. Nevertheless, it is still interesting to compare the performance of the "iterated ants" algorithms to the IG ones since both use different rules for constructing complete candidate solutions. Exemplary results for this comparison with plots of the development of the solution quality over computation time are in Figure 4.

The overall best performing variant is IG-LS-CVSPTW. Across the 25 test instances, it is statistically better than "iterated ants" with local search on 12 instances and on none statistically worse. The situation is less clear, when considering the variants without local search; here on some instances IG-CVSPTW performs better than the "iterated ants" variant (such as the KV instance), while the opposite is true on others (such as on the largest instance tested here, EW).

7   Discussion

A direct comparison of IG-LS-CVSPTW to the performance of the best ILP-based approach in [1] would certainly be interesting. However, a direct comparison is, at least with the current heuristic algorithm code, not straightforward because of some minor differences in the constraints considered. In fact, in the greedy heuristics, a maximum cycle length of 24 hours is imposed for the schedule of each locomotive, which is not done in the ILP formulation: in the ILP formulation a schedule cycle of a locomotive can be an arbitrary multiple of 24 hours. Hence, the heuristic solutions generated within the 24 hour limit for each locomotive is a subset of the ones considered by the ILP approach. Anyway, the solutions generated by the heuristics are feasible for the ILP formulation, that is, despite this difference, the heuristic solutions are valid upper bounds for the ILP. The difficulty in comparing our solutions to the ILP ones is that the ILP formulation allows possibly much better solutions to be reached. If we anyway compare the solutions, then we can see that for the same instance–time window

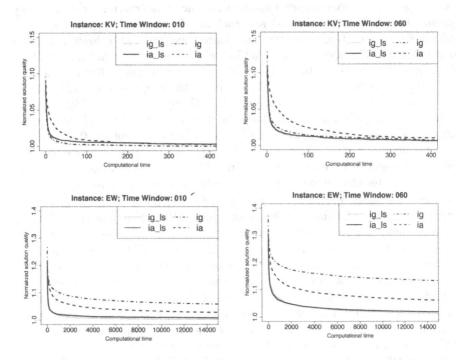

Fig. 4. Development of the solution quality over time for iterated greedy, iterated greedy plus local search, iterated ants and iterated ants plus local search for instances KV-10 (top left), KV-60 (top right), EW-10 (bottom left), and EW-60 (bottom right)

combination, the IG-LS-CVSPTW algorithm typically finds solutions which use the same number of locomotives or one or two more than in available optimal solutions. Additionally, our heuristics have the advantage that they are applicable to very large instances, which are beyond the reach of the ILP-based approach. In fact, already for the largest instance for which the ILP solver could still deliver solutions in [1] (instance EW with time windows ±30), improved upper bounds could be found by IG-LS-CVSPTW.

Our heuristic algorithms can generate good quality in relatively short computation times on the instances tested here, say a few minutes for the largest instance tested here. Given more computation time, IG-CVSPTW and IG-LS-CVSPTW can use this time effectively and further improve the quality of the solutions over time. On the largest instance tested here, computation times up to four hours have been considered to test the limiting behavior of the heuristics (see, for example, the results given in Figure 4 on instance EW). The computational effort is, however, still lower than the one given to the ILP approach, which was run for one CPU hour in parallel on eight cores of a similar machine as ours (Cplex 10 makes effective use of multiple cores by parallelizing the branch-and-bound tree computations.). In addition, there are still a number of possibilities for improving the speed of the heuristic algorithms, ranging from the adoption of pre-processing techniques to reduce instance size (which is actually done for the ILP model) to more fine-tuned implementations.

Concerning the results for the iterated ants, one may wonder, whether the setting of $\alpha = 0$ is an artifact of the tuning and whether on larger instances the usage of pheromone trails would result into better performance. (Note that the KV and EW instances are in part much larger than the ones used for tuning.) To test this, we took the second ranked algorithm configuration from the race, which had for both cases–with and without local search–settings of α around one. However, on the two largest instances no significant improvements by using pheromone trails in the solution construction have been identified.

The negative results of the tentative extension of IG to iterated ants allows two different interpretations. On one side, it does not exclude that an iterated ants approach or another population-based extension may further improve performance. For example, different ways of defining the meaning of the pheromone trails may be tested, such as associating locomotive types to trips. However, significant further developments and tests would have to be done, time which could be also used to further improve the simpler algorithm, for example, by using more elaborate construction heuristics or improved local search algorithms. On the other side, these tests are also a confirmation that conceptually rather simple algorithms are a good means to improve the performance of basic heuristics especially in the context of real-world problems. In this sense, the results here give also an example of how a bottom-up development of SLS algorithms, which adds algorithm features in a step-by-step manner, is a viable approach to obtain high performing yet conceptually simple algorithms.

8   Conclusions

In this paper we have developed a high performing stochastic local search algorithm for a freight train scheduling problem arising in the strategical planning of Deutsche Bahn AG. In particular, we have shown that a combination of an iterated greedy heuristic with a simple local search algorithm yields very promising performance. With this algorithm we now can obtain high quality solutions to large problem instances, for which an approach based on a commercial solver ILP solver is not effective anymore.

There are a number of directions in which this research could be extended. A first is certainly the application and comparison of the iterated greedy and iterated ants algorithms on very large instances. For tackling large instances, computation time reductions may be obtained by adopting a pre-processing phase to reduce the effective instance size tackled (this was also indicated by initial tests). Another attractive possibility is to consider hybrids between the exact solution methods and the iterated greedy algorithms. One way to do this is by exploiting the very good performance of the commercial solver for small sized instances: One may use the iterated greedy algorithm to define partial solutions and compute their optimal extensions to complete ones by an ILP solver.

Acknowledgments. Zhi Yuan acknowledges support from COMP²SYS, a Marie Curie Early Stage Research Training Site funded by the European Community's Sixth Framework Programme and the ANTS project, an *Action de Recherche Concerteé* funded by the Scientific Research Directorate of the French Community of Belgium. Thomas Stützle acknowledges support from the Belgian F.R.S.-FNRS of which he is a Research Associate. Henning Homfeld acknowledges support from the German Ministry for Science BMBF, program "Math and Industry". The authors would like to thank also Dr. Gerald Pfau, Andreas Ginkel, and Jörg Wolfner from Deutsche Bahn AG for providing the problem and for fruitful and productive discussions.

References

1. Fügenschuh, A., Homfeld, H., Huck, A., Martin, A., Yuan, Z.: Scheduling locomotives and car transfers in freight transport (preprint),
   http://www.hausdorff-research-institute.uni-bonn.de/
   files/preprints/2006transsci.pdf (submitted to Transportation Science)
2. ILOG Ltd.: ILOG Cplex 10 Solver Suite. Technical report, ILOG Cplex Division, 889 Alder Avenue, Suite 200, Incline Village, NV 89451, USA (2006)
3. Ruiz, R., Stützle, T.: A simple and effective iterated greedy algorithm for the permutation flowshop scheduling problem. European Journal of Operational Research 177(3), 2033–2049 (2007)
4. Balaprakash, P., Birattari, M., Stützle, T.: Improvement strategies for the f-race algorithm: Sampling design and iterative refinement. In: Bartz-Beielstein, T., Blesa Aguilera, M.J., Blum, C., Naujoks, B., Roli, A., Rudolph, G., Sampels, M. (eds.) HM 2007. LNCS, vol. 4771, pp. 108–122. Springer, Heidelberg (2007)
5. Fügenschuh, A.: Parametrized Greedy Heuristics in Theory and Practice. In: Blesa, M.J., Blum, C., Roli, A., Sampels, M. (eds.) HM 2005. LNCS, vol. 3636, pp. 21–31. Springer, Heidelberg (2005)
6. Charon, I., Hudry, O.: The noising method: A new method for combinatorial optimization. Operations Research Letters 14(3), 133–137 (1993)
7. Hoos, H.H., Stützle, T.: Stochastic Local Search—Foundations and Applications. Morgan Kaufmann Publishers/Elsevier, San Francisco (2004)
8. Wiesemann, W., Stützle, T.: Iterated ants: An experimental study for the quadratic assignment problem. In: Dorigo, M., Gambardella, L.M., Birattari, M., Martinoli, A., Poli, R., Stützle, T. (eds.) ANTS 2006. LNCS, vol. 4150, pp. 179–190. Springer, Heidelberg (2006)
9. Stützle, T., Hoos, H.: $\mathcal{MAX}$–$\mathcal{MIN}$ Ant System and local search for combinatorial optimization problems. In: Voss, S., Martello, S., Osman, I.H., Roucairol, C. (eds.) Meta-Heuristics: Advances and Trends in Local Search Paradigms for Optimization, pp. 137–154. Kluwer Academic Publishers, Dordrecht (1999)

On the Integration of a TSP Heuristic into an EA for the Bi-objective Ring Star Problem

Arnaud Liefooghe[1], Laetitia Jourdan[1], Nicolas Jozefowiez[2,3], and El-Ghazali Talbi[1]

[1] LIFL – CNRS – INRIA Lille-Nord Europe,
Université des Sciences et Technologies de Lille,
Parc Scientifique de la Haute Borne, 40 av. Halley, 59650 Villeneuve d'Ascq, France
{Arnaud.Liefooghe,Laetitia.Jourdan,El-Ghazali.Talbi}@lifl.fr
[2] LAAS – CNRS, Université de Toulouse,
7 av. du Colonel Roche, F-31077 Toulouse, France
Nicolas.Jozefowiez@laas.fr
[3] Université de Toulouse, INSA, France

Abstract. This paper discusses a new hybrid solution method for a bi-objective routing problem, namely the bi-objective ring star problem. The bi-objective ring star problem is a generalization of the ring star problem in which the assignment cost has been dissociated from the cost of visiting a subset of nodes. Here, we investigate the possible contribution of incorporating specialized TSP heuristics into a multi-objective evolutionary algorithm. Experiments show that the use of this hybridization scheme allows a strict improvement of the generated sets of non-dominated solutions.

1 Introduction

The purpose of the Bi-objective Ring Star Problem (B-RSP) is to locate an elementary cycle, the so-called *ring*, on a subset of nodes of a graph while optimizing two conflicting costs. First is the minimization of a *ring cost*, proportional to the length of the cycle. Then, nodes that do not belong to the ring are all assigned to visited ones so that the associated cost is minimal. The resulting *assignment cost* is the second objective to be minimized. In spite of its natural bi-objective formulation, this problem is generally investigated in a single-objective way, either where both costs are combined [12] or where the assignment cost is treated as a constraint [13]. Note that both versions of the problem have also been heuristically solved in [16,18]. As pointed out in [10], a large number of routing problems are formulated as multi-objective optimization problems, and according to the same paper, the B-RSP is a generalization of a mono-objective problem. In [15], different multi-objective evolutionary algorithms have been proposed for the B-RSP. Although the approaches were already encouraging, even compared to state-of-the-art mono-objective methods, a few improvement points can be identified. First, a lack of efficiency has been detected on the ring cost throughout the output solutions. Second, the population initialization strategy used within

M.J. Blesa et al. (Eds.): HM 2008, LNCS 5296, pp. 117–130, 2008.
© Springer-Verlag Berlin Heidelberg 2008

all the search methods was a bit rudimentary, each initial solution having approximately half of its nodes in the cycle. The challenge is then to overcome the identified problems in order to improve the efficiency of those search methods.

An interesting property of the problem under consideration is that, given a fixed set of visited nodes, the related assignment cost is always optimal. It is not the case for the ring cost, for which a classical Traveling Salesman Problem (TSP) is still to be solved among the set of nodes that belong to the ring. Then, once is decided which nodes are visited or not, an objective function is much more difficult to optimize. However, a large number of efficient heuristic methods has been proposed for the TSP. In this paper, our aim is to present a hybrid metaheuristic combining a multi-objective evolutionary algorithm and a problem-specific heuristic, initially designed for the TSP. Approaches where a TSP heuristic is successfully integrated into a multi-objective evolutionary algorithm can, for instance, be found in [8,9].

The reminder of the paper is organized as follows. In Section 2, we give the necessary background for multi-objective optimization, we introduce the B-RSP and we present a heuristic devoted to the TSP. The hybrid metaheuristic proposed to solve the B-RSP is detailed in Section 3. In Section 4, computational experiments are conducted. At last, conclusions and perspectives are drawn in the last section.

2   Background

In this section, we first discuss multi-objective optimization and define some related concepts. Then, we present the bi-objective ring star problem in details and we introduce a heuristic devoted to the traveling salesman problem.

2.1   Multi-Objective Optimization

A general *Multi-objective Optimization Problem* (MOP) can be defined by a set of $n \geq 2$ objective functions $f_1, f_2, \ldots, f_n$; a set of feasible solutions in the *decision space*, denoted by X; and a set of feasible points in the *objective space*, denoted by Z. Each function can be either minimized or maximized, but we here assume that all n objective functions are to be minimized. To each decision vector $x \in X$ is assigned exactly one objective vector $z \in Z$ on the basis of a vector function $f : X \to Z$ with $z = f(x) = (f_1(x), f_2(x), \ldots, f_n(x))$.

Definition 1. *An objective vector $z \in Z$ weakly dominates another objective vector $z' \in Z$ if and only if $\forall i \in [1..n]$, $z_i \leq z'_i$.*

Definition 2. *An objective vector $z \in Z$ dominates another objective vector $z' \in Z$ if and only if $\forall i \in [1..n]$, $z_i \leq z'_i$ and $\exists j \in [1..n]$ such as $z_j < z'_j$.*

Definition 3. *An objective vector $z \in Z$ is non-dominated if and only if there does not exist another objective vector $z' \in Z$ such that z' dominates z.*

A solution $x \in X$ is said to be *efficient* (or *non-dominated*) if $f(x)$ is non-dominated. The set of all efficient solutions is the *efficient set*, denoted by X_E. The set of all non-dominated vectors is the *non-dominated front* (or the *trade-off surface*), denoted by Z_N. A possible approach to solve a MOP consists of finding or approximating a minimal set of efficient solutions, *i.e.* one solution $x \in X_E$ for each non-dominated point $z \in Z_N$ such as $f(x) = z$ (in case multiple solutions map to the same non-dominated vector). Evolutionary algorithms are commonly used to this end as they naturally find multiple and well-spread non-dominated solutions in a single simulation run. The reader could refer to [3,4] for more details about evolutionary multi-objective optimization.

2.2 The Bi-objective Ring Star Problem

The *Bi-objective Ring Star Problem* (B-RSP) can be described as follows. Let $G = (V, E, A)$ be a complete mixed graph where $V = \{v_1, v_2, \ldots, v_n\}$ is a set of vertexes, $E = \{[v_i, v_j] | v_i, v_j \in V, i < j\}$ is a set of edges, and $A = \{(v_i, v_j) | v_i, v_j \in V\}$ is a set of arcs. Vertex v_1 is the depot. To each edge $[v_i, v_j] \in E$ we assign a non-negative *ring cost* c_{ij}, and to each arc $(v_i, v_j) \in A$ we assign a non-negative *assignment cost* d_{ij}. The B-RSP consists of locating a simple cycle through a subset of nodes $V' \subset V$ (with $v_1 \in V'$) while (i) minimizing the sum of the ring costs related to all edges that belong to the cycle, and (ii) minimizing the sum of the assignment costs of arcs directed from every non-visited node to a visited one so that the associated cost is minimum. An example of solution is given in Figure 1, where solid lines represent edges that belong to the ring and dashed lines represent arcs of the assignments.

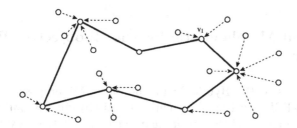

Fig. 1. An example of solution for the ring star problem

The first objective is called the *ring cost* and is defined as:

$$\sum_{[v_i, v_j] \in E} c_{ij} b_{ij} \ , \tag{1}$$

where b_{ij} is a binary variable equal to 1 if and only if the edge $[v_i, v_j]$ belongs to the cycle. The second objective, the *assignment cost*, can be computed as follows:

$$\sum_{v_i \in V \setminus V'} \min_{v_j \in V'} d_{ij} \ . \tag{2}$$

Let us remark that these two objectives are comparable only if we assume that the ring cost and the assignment cost are commensurate one to another, what is rarely the case in practice. Furthermore, the fact of privileging a cost compared to the other is closely related to the decision-maker preferences. However, the B-RSP is an NP-hard combinatorial problem since the particular case of visiting the whole set of nodes is equivalent to a traditional Traveling Salesman Problem (TSP).

2.3   GENIUS, a TSP Heuristic

A specificity of the B-RSP is that many TSP generally need to be solved. An effective TSP heuristic method is thus rather appreciated in order to improve the ring cost of a solution. There exists a large range of heuristics that are devoted to the TSP. One of them is GENIUS, proposed by Gendreau et al. [6]. Briefly, GE-NIUS contains a tour construction phase, called GENI, and a postoptimization phase, called US. Starting with three arbitrary nodes, GENI inserts, at each iteration, an unrouted node between two of its p closest neighbors on the partially constructed tour, where p is a user-controlled parameter. When inserting the vertex, GENI also performs a local reoptimization of the tour. Once a complete tour has been built, the US postoptimization procedure is repeatedly applied to the tour until no further improvement is possible. During this procedure, nodes are successively removed from the tour, and then reinserted, according to the same rules used in the tour construction phase. The use of GENIUS can be seen as a black-box mechanism integrated into the hybrid metaheuristic presented in the next section, and could practically be replaced by another TSP heuristic.

3   A Hybrid Metaheuristic for the Bi-objective Ring Star Problem

The main process of the Hybrid Metaheuristic (HM) proposed in the paper to solve the B-RSP consists of an elitist multi-objective Evolutionary Algorithm (EA). A first hybridization mechanism arises at the very beginning of the HM, as the initial population is built thanks to a problem-specific heuristic. This initial population is used as a starting point of the EA, so that both methods cooperate in pipeline way. Second, an additional hybridization scheme conditionally appears at every generation of the EA, where the ring cost of each population member is attempted to be improved thanks to a TSP heuristic. Finally, the EA is itself hybrid, as it is divided into two different phases. Those ones differ the one from the other at the selection and the replacement steps of the EA. During both phases, a secondary population, the so-called archive, is used to store every potentially efficient solutions found so far. The first phase is compound of an elitist selection step where parent individuals are all selected from the archive only. The replacement step is a generational one, *i.e.* the parent population is replaced by the offspring one. This phase corresponds to the *Simple*

Elitist Evolutionary Algorithm (SEEA) introduced in [15]. The main particularity of SEEA is that no fitness assignment scheme is required, the population being the only problem-independent parameter. The second phase is founded on the *Indicator-Based Evolutionary Algorithm* proposed by Zitzler and Künzli [21]. The fitness assignment scheme of IBEA is based on a pairwise comparison of population items by using a binary quality indicator I. Several indicators can be used for such a purpose [21], and we here choose to use the binary additive ϵ-indicator ($I_{\epsilon+}$) proposed in [23]. $I_{\epsilon+}$ gives the minimum factor by which a non-dominated set A has to be translated in the objective space to weakly dominate a non-dominated set B. The selection scheme for reproduction is a binary tournament between randomly chosen individuals. The replacement strategy consists of deleting, one-by-one, the worst individuals, and in updating the fitness values of the remaining solutions each time there is a deletion; this is continued until the required population size is reached. The first phase of the EA will allow to find a rough approximation of the efficient set in a very short amount of time whereas the second phase will rather be devoted to improve this set in a more intensive way. The transition from Phase 1 to Phase 2 will occur as soon as the archive of non-dominated solutions does not improve enough with regards to the search scenario. The main steps of our HM are the following ones:

1. **Initialization.** Generate an initial population P of size N (see Section 3.2); generate an efficient set approximation A with the non-dominated individuals contained in P; create an empty offspring population P'.
2. **Selection.** Repeat until $|P'| = N$:
   (Phase 1) Randomly select an individual from A and add it to the offspring population P'.
   (Phase 2) Select an individual thanks to a binary tournament selection on P and add it to the offspring population P'.
3. **Recombination.** Apply a recombination operator to pairs of individuals contained in P' with a given probability p_r (see Section 3.3).
4. **Mutation.** Apply a mutation operator to individuals contained in P' with a given probability p_m (see Section 3.4).
5. **Fitness assignment.**
   (Phase 1) $\emptyset$.
   (Phase 2) Calculate fitness values of any individual x contained in $P \cup P'$; i.e. $F(x) \leftarrow \sum_{x' \in (P \cup P') \setminus \{x\}} -e^{-I(\{x'\}, \{x\})/\kappa}$, where $\kappa > 0$ is a scaling factor.
6. **Replacement.**
   (Phase 1) $P \leftarrow P'$; $P' \leftarrow \emptyset$.
   (Phase 2) $P \leftarrow P \cup P'$; $P' \leftarrow \emptyset$. Iterate the following steps until the size of the population P does not exceed N:
   - Choose an individual $x^\star \in P$ with the smallest fitness value; i.e. $F(x^\star) \leq F(x)$ for all $x \in P$.
   - Remove $x^\star$ from P.
   - Update the fitness values of the individuals remaining in P; i.e. $F(x) \leftarrow F(x) + e^{-I(\{x^\star\}, \{x\})/\kappa}$ for all $x \in P$.

7. **Elitism.** $A \leftarrow$ non-dominated individuals of $A \cup P$.
8. **Improvement.** If a given condition is satisfied, apply an improvement procedure on any individual contained in P (see Section 3.5).
9. **Termination.** If a stopping criteria is satisfied return A, else go to Step 2.

The principle of the HM is illustrated in Figure 2. According to the taxonomy of hybrid metaheuristics proposed in [20], the HM proposed in this paper can be classified on the *high-level relay hybrid* class, where self-contained heuristics are executed in sequence. The problem-specific components are explained in details below.

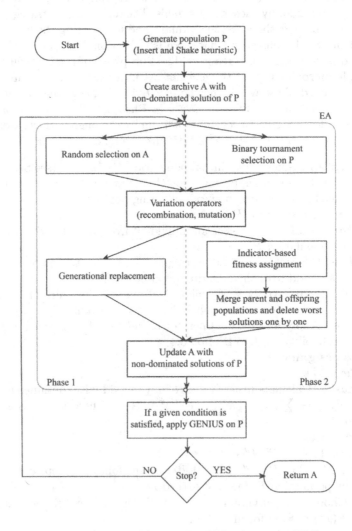

Fig. 2. Flowchart of the Hybrid Metaheuristic (HM)

Vertex	v_1	v_2	v_3	v_4	v_5	v_6	v_7	v_8	v_9	v_{10}
Random key	0	0.7	-	0.3	-	0.8	0.2	-	0.5	-

Fig. 3. A RSP solution represented by random keys

3.1   Solution Encoding

The representation of a B-RSP solution is based on the random keys mechanism proposed by Bean [1]. A *random key* $k_i \in [0, 1[$ is assigned to every node v_i that belongs to the ring. A special value is assigned to unvisited nodes. Thus, the ring route associated to a solution corresponds to the nodes read according to their random keys in the increasing order; *i.e.* if $k_i < k_j$, then v_j comes after v_i. A possible representation for the cycle $(v_1, v_7, v_4, v_9, v_2, v_6)$ is given in Figure 3. Nodes v_3, v_5, v_8 and v_{10} are assigned to a visited node in such a way that the associated assignment cost is minimum.

3.2   Population Initialization

An initial population of N individuals is built by means of repeatedly solving a mono-objective problem closely linked to the B-RSP. This problem, that will be denoted by Ring Cost Constrained RSP (RCC-RSP), consists of minimizing the assignment cost only, while satisfying an upper bound on the ring cost. It is obtained by removing the ring cost from the set of objective functions of the B-RSP, and by adding a new constraint stipulating that the ring cost cannot exceed a given limit $c_{\max}$. A search mechanism is iterated with distinct $c_{\max}$ values such that the set of resulting problems corresponds to different part of the objective space.

In order to approximately solve a given RCC-RSP, we use the *Insert and Shake Heuristic* (ISH), initially proposed by Gendreau et al. [7] for a single-objective routing problem called the selective TSP. This method combines a TSP tour extension heuristic described in Rosenkrantz et al. [19] and the GENIUS procedure described in Section 2.3. ISH gradually extends a tour T until no other node can be added without violating a given ring cost limit $c_{\max}$. At a given step, the non-visited node to be inserted in T is chosen so that the ratio between its current assignment cost and the increment on the global ring cost after its insertion is minimum. Then, GENIUS is applied in an attempt to obtain a better cycle on the nodes in T. If GENIUS fails, the procedure terminates. Otherwise, more node insertions are attempted and the process is repeated. The steps to build an initial population of solutions are the following ones:

1. GENIUS is applied to find a cycle containing all nodes in V, and this solution is included in the population. Let $c^\star$ be the ring cost of this solution. Set $\alpha \leftarrow \frac{c^\star}{N-1}$, $c_{\max} \leftarrow c^\star - \alpha$ and $i \leftarrow 1$.
2. If $i > N$, stop. Otherwise, generate a RCC-RSP solution by means of ISH and insert this solution into the population.
3. Set $c_{\max} \leftarrow c_{\max} - \alpha$, and $i \leftarrow i + 1$. Go to Step 2.

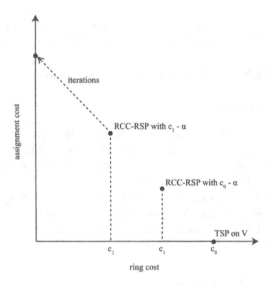

Fig. 4. Illustration of the population initialization heuristic

This initialization strategy is illustrated in Figure 4. Thanks to this heuristic, the starting set of solutions will already be both (i) quite efficient, and (ii) well-spread on the objective space.

3.3   Recombination Operator

The recombination operator is a quadratic crossover closely related to the one proposed in [18]. Two randomly selected solutions x_1 and x_2 are first divided according to a particular position. Then, the first part of x_1 is combined with the second part of x_2 to build a first offspring, and the first part of x_2 is combined with the second part of x_1 to build a second offspring. Every node retains its random key so that it enables an easy reconstruction of the new individuals. Thanks to the random keys encoding mechanism, solutions having different ring sizes can easily be recombined, even if the initial ring structures are generally broken in the offspring solutions.

3.4   Mutation Operator

The mutation operator designed for the problem under consideration consists of the following strategy. A node $v^\star \in V \setminus \{v_1\}$ is selected at random. Therefore, two cases may arise. First, if $v^\star$ belongs to the ring, it is removed and then belongs to the set of unvisited nodes. Second, if $v^\star$ does not belong to the cycle, it is added. The position to insert $v^\star$ is chosen so that the increment on the ring cost is minimum.

3.5   Improvement Procedure

The improvement procedure consists of improving the ring cost of the current population by applying GENIUS on any of its members. The main issue is now to determine when this heuristic might start in order to find the good trade-off between the efficiency and the effectiveness of GENIUS. To do so, we decide to launch it only if no more than n solutions per iteration have been included in the archive during the last M consecutive generations. Hence, we hope that the improved population will produce new non-dominated solutions and will help to build more interesting individuals in the future steps of the HM.

4   Computational Experiments

In order to assess the effectiveness of our method, we will measure its performance in comparison to the ACS method proposed in [15]. The latter is an auto-adaptive method based on a simple elitist evolutionary algorithm and a population-based local search. It has been shown to be particularly efficient to solve the RSP as a bi-objective problem. To quantity the impact of GENIUS on our HM, we also implemented a more basic version in which GENIUS is not involved, neither in the improvement step nor in the initialization step. This other hybrid metaheuristic will be denoted by HM2 in the remainder of the paper. All the algorithms have been implemented under the ParadisEO-MOEO library[1] [14] and share the same base components for a fair comparison between them. Computational runs were performed under Linux on an Intel Core 2 Duo 6600 (2×2.40 GHz) machine, with 2 GB RAM.

4.1   Performance Assessment

Experiments have been conducted on a set of benchmark test instances taken from the TSPLIB[2] [17]. These instances involve between 51 and 264 nodes. The number at the end of an instance's name represents the number of nodes for the instance under consideration. Let l_{ij} denote the distance between two nodes v_i and v_j of a TSPLIB file. Then, the ring cost c_{ij} and the assignment cost d_{ij} have both been set to l_{ij} for every pair of nodes v_i and v_j.

For each search method, a set of 20 runs per instance has been performed. In order to evaluate the quality of the non-dominated front approximations, we follow the protocol given by Knowles et al. [11]. For a given instance, we first compute a reference set $Z_N^\star$ of non-dominated points extracted from the union of all the fronts we obtained during our experiments and the best non-dominated set taken from [15][3]. Second, we define a point $z^{max} = (z_1^{max}, z_2^{max})$, where z_1^{max} (respectively z_2^{max}) denotes the upper bound of the first (respectively second) objective in the whole non-dominated front approximations. Then, to

---

[1] ParadisEO is available at http://paradiseo.gforge.inria.fr

[2] http://www.iwr.uni-heidelberg.de/groups/comopt/software/TSPLIB95/

[3] These results are available at http://www.lifl.fr/~liefooga/rsp/

measure the quality of an output set A in comparison to $Z_N^\star$, we compute the difference between these two sets by using the unary hypervolume metric [22], $(1.05 \times z_1^{max}, 1.05 \times z_2^{max})$ being the reference point. The hypervolume difference indicator (I_H^-) computes the portion of the objective space that is weakly dominated by $Z_N^\star$ and not by A. The more this measure is close to 0, the better is the approximation A. Furthermore, we also consider one of the ϵ-indicators proposed in [23]. The unary additive ϵ-indicator ($I_{\epsilon+}^1$) gives the minimum factor by which an approximation A has to be translated in the objective space to weakly dominate the reference set $Z_N^\star$. As a consequence, for each test instance, we obtain 20 I_H^- measures and 20 $I_{\epsilon+}$ measures, corresponding to the 20 runs, per algorithm. As suggested by Knowles et al. [11], once all these values are computed, we perform a statistical analysis on pairs of optimization methods for a comparison on a specific test instance. To this end, we use the Mann-Whitney statistical test as described in [11], with a p-value lower than 5%. Hence, for a specific test instance, and according to the p-value and to the metric under consideration, this statistical test reveals if the sample of approximation sets obtained by a given search method is significantly better than the one obtained by another search method, or if there is no significant difference between both. Note that all the performance assessment procedures have been achieved using the performance assessment tool suite provided in PISA[4] [2].

4.2   Parameter Setting

For each investigated metaheuristic, the search process stops after a certain amount of run time. As shown in Table 1, this stopping criteria has been arbitrary set according to the size of the problem instance to be solved. Next, the population size N is set to 100; the recombination probability p_r is set to 0.25 and the mutation probability p_m is set to 1.0. Following [21], the scaling factor κ is set to 0.05. The improvement procedure of HM is launched only if the number of elements received by the archive is less than 1.0% of its current size for $|V|$ consecutive iterations, where $|V|$ is the number of nodes for the instance under consideration. Finally, the GENIUS parameter p is set to 7.

Table 1. Stopping criteria: running time

Instance	Running time	Instance	Running time
eil51	20"	kroA150	10'
st70	1'	kroA200	20'
kroA100	2'	pr264	30'
bier127	5'		

---

[4] The package is available at http://www.tik.ee.ethz.ch/pisa/assessment.html

Table 2. Comparison of HM, HM2 and ACS [15] according to the I_H^- and the $I_{\epsilon+}$ metrics by using a Mann-Whitney statistical test with a p-value of 5%. For the metric under consideration, either the results of the algorithm located at a specific row are significantly better than those of the algorithm located at a specific column ($\succ$), either they are worse ($\prec$), or there is no significant difference between both ($\equiv$).

		I_H^-			$I_{\epsilon+}$		
		HM	HM2	ACS	HM	HM2	ACS
eil51	HM	-	$\succ$	$\succ$	-	$\succ$	$\succ$
	HM2	$\prec$	-	$\equiv$	$\prec$	-	$\equiv$
	ACS	$\prec$	$\equiv$	-	$\prec$	$\equiv$	-
st70	HM	-	$\succ$	$\succ$	-	$\succ$	$\succ$
	HM2	$\prec$	-	$\equiv$	$\prec$	-	$\equiv$
	ACS	$\prec$	$\equiv$	-	$\prec$	$\equiv$	-
kroA100	HM	-	$\succ$	$\succ$	-	$\succ$	$\succ$
	HM2	$\prec$	-	$\prec$	$\prec$	-	$\prec$
	ACS	$\prec$	$\succ$	-	$\prec$	$\succ$	-
bier127	HM	-	$\succ$	$\succ$	-	$\succ$	$\succ$
	HM2	$\prec$	-	$\succ$	$\prec$	-	$\succ$
	ACS	$\prec$	$\prec$	-	$\prec$	$\prec$	-
kroA150	HM	-	$\succ$	$\succ$	-	$\succ$	$\succ$
	HM2	$\prec$	-	$\equiv$	$\prec$	-	$\prec$
	ACS	$\prec$	$\equiv$	-	$\prec$	$\succ$	-
kroA200	HM	-	$\succ$	$\succ$	-	$\succ$	$\succ$
	HM2	$\prec$	-	$\equiv$	$\prec$	-	$\prec$
	ACS	$\prec$	$\equiv$	-	$\prec$	$\succ$	-
pr264	HM	-	$\succ$	$\succ$	-	$\succ$	$\succ$
	HM2	$\prec$	-	$\equiv$	$\prec$	-	$\prec$
	ACS	$\prec$	$\equiv$	-	$\prec$	$\succ$	-

4.3 Results and Discussion

First of all, note that we initially experimented some algorithm versions where only the first or the second phase of the EA is involved, with and without GE-NIUS. But the resulting metaheuristics turned out to be significantly outperformed by HM and HM2. The comparison of results obtained by HM, HM2 and ACS are presented in Table 2. According to both indicators (I_H^- and $I_{\epsilon+}$), HM is statistically better than any other search methods on every instance we investigated. Besides, the difference between HM2 and ACS is often not significant according to the I_H^- metric, whereas ACS generally outperforms HM2 according to the $I_{\epsilon+}$ metric. The only instance for which HM2 performs statistically higher than ACS is the *bier127* instance, where it obtains better values for both metrics. In order to study on which part of the trade-off surface the differences between HM, HM2 and ACS appear, examples of empirical attainment functions [5] are given in Figure 5 and Figure 6 for the *bier127* instance. They represent the limit of the objective space that is attained by at least 90% of the runs for every search method. For the instance under consideration, we can see that HM seems to be more capable of finding solutions having both a large number of visited nodes

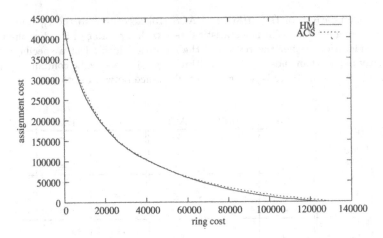

Fig. 5. 90%-attainment surface plot obtained by the approximation sets found by HM and ACS [15] for the *bier127* test instance

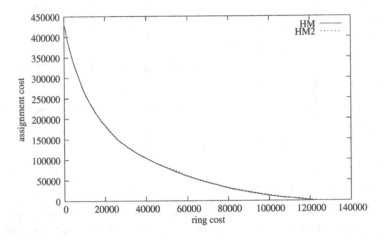

Fig. 6. 90%-attainment surface plot obtained by the approximation sets found by HM and HM2 for the *bier127* test instance

and a good ring cost. Thus, the superiority of HM relatively to HM2 reveals the benefit of integrating a TSP heuristic, here symbolized by GENIUS, into our EA for the problem to be solved. Moreover, despite its relative simplicity in comparison to ACS, the HM is quite effective to solve the B-RSP, especially to find solutions having a low ring cost. This indicates that the hybridization scheme largely improves the method and reveals that the HM introduced in this paper outperforms the metaheuristics proposed so far to solve the RSP as a bi-objective optimization problem.

5   Conclusion and Perspectives

In this paper, a new hybrid metaheuristic has been proposed to approximate the efficient set of a multi-objective routing problem called the bi-objective ring star problem. This problem is commonly investigated in a single-objective way, either where both objectives are aggregated, or where one objective is regarded as a constraint. However, within the frame of the ring star problem, many traveling salesman problems generally need to be solved. The purpose of the hybrid metaheuristic proposed here is then to integrate a heuristic algorithm for the traveling salesman problem, namely GENIUS, into a multi-objective evolutionary algorithm to solve the bi-objective ring star problem as a whole. The hybrid search method starts with a problem-specific heuristic to generate an initial set of solutions, and continues with a two-phase elitist evolutionary algorithm hybridized to the GENIUS heuristic. The latter is launched to intensify the search in an auto-adaptive manner, according to the convergence scenario of the main process. Experiments were conducted on a set of benchmark test instances, and validated the contribution of the traveling salesman problem heuristic into the hybrid method. They also reveal that the metaheuristic proposed in the paper largely outperforms our previous investigations for resolving the bi-objective ring star problem. As a next step, we will experiment other strategies to scale the GENIUS application factor in order to study the influence of this parameter on the global performance of our method. We will also try to replace GENIUS by other kinds of traveling salesman problem heuristics, or even exact methods, within our hybrid metaheuristic to assert the genericity of our method. Moreover, it could be interesting to design a more universal model of hybridization to solve multi-objective optimization problems, where both problem-specific and meta methods could be integrated.

References

1. Bean, J.: Genetic algorithms and random keys for sequencing and optimization. ORSA Journal on Computing 6(2), 154–160 (1994)
2. Bleuler, S., Laumanns, M., Thiele, L., Zitzler, E.: PISA — a platform and programming language independent interface for search algorithms. In: Fonseca, C.M., Fleming, P.J., Zitzler, E., Deb, K., Thiele, L. (eds.) EMO 2003. LNCS, vol. 2632, pp. 494–508. Springer, Heidelberg (2003)
3. Coello Coello, C.A., Van Veldhuizen, D.A., Lamont, G.B.: Evolutionary Algorithms for Solving Multi-Objective Problems. Kluwer Academic Publishers, Boston (2002)
4. Deb, K.: Multi-Objective Optimization using Evolutionary Algorithms. John Wiley & Sons, Chichester (2001)
5. Fonseca, C.M., Grunert da Fonseca, V., Paquete, L.: Exploring the performance of stochastic multiobjective optimisers with the second-order attainment function. In: Coello Coello, C.A., Hernández Aguirre, A., Zitzler, E. (eds.) EMO 2005. LNCS, vol. 3410, pp. 250–264. Springer, Heidelberg (2005)
6. Gendreau, M., Hertz, A., Laporte, G.: New insertion and postoptimization procedures for the traveling salesman problem. Operations Research 40(6), 1086–1094 (1992)

7. Gendreau, M., Laporte, G., Semet, F.: A tabu search heuristic for the undirected selective travelling salesman problem. European Journal of Operational Research 106, 539–545 (1998)
8. Jozefowiez, N., Glover, F., Laguna, M.: Multi-objective meta-heuristics for the traveling salesman problem with profits. Journal of Mathematical Modelling and Algorithms 7(2), 177–195 (2008)
9. Jozefowiez, N., Semet, F., Talbi, E.-G.: The bi-objective covering tour problem. Computers and Operations Research 34(7), 1929–1942 (2007)
10. Jozefowiez, N., Semet, F., Talbi, E.-G.: Multi-objective vehicle routing problems. European Journal of Operational Research 189(2), 293–309 (2008)
11. Knowles, J., Thiele, L., Zitzler, E.: A tutorial on the performance assessment of stochastic multiobjective optimizers. Technical report, Computer Engineering and Networks Laboratory (TIK), ETH Zurich, Switzerland (revised version) (2006)
12. Labbé, M., Laporte, G., Rodríguez Martín, I., Salazar González, J.J.: The ring star problem: Polyhedral analysis and exact algorithm. Networks 43, 177–189 (2004)
13. Labbé, M., Laporte, G., Rodríguez Martín, I., Salazar González, J.J.: Locating median cycles in networks. European Journal of Operational Research 160(2), 457–470 (2005)
14. Liefooghe, A., Basseur, M., Jourdan, L., Talbi, E.-G.: ParadisEO-MOEO: A framework for evolutionary multi-objective optimization. In: Obayashi, S., Deb, K., Poloni, C., Hiroyasu, T., Murata, T. (eds.) EMO 2007. LNCS, vol. 4403, pp. 386–400. Springer, Heidelberg (2007)
15. Liefooghe, A., Jourdan, L., Talbi, E.-G.: Metaheuristics and their hybridization to solve the bi-objective ring star problem: a comparative study. Working paper RR-6515, Institut National de Recherche en Informatique et Automatique (INRIA) (2008)
16. Moreno Pérez, J.A., Moreno-Vega, J.M., Rodríguez Martín, I.: Variable neighborhood tabu search and its application to the median cycle problem. European Journal of Operations Research 151(2), 365–378 (2003)
17. Reinelt, G.: TSPLIB – A traveling salesman problem library. ORSA Journal on Computing 3(4), 376–384 (1991)
18. Renaud, J., Boctor, F.F., Laporte, G.: Efficient heuristics for median cycle problems. Journal of the Operational Research Society 55(2), 179–186 (2004)
19. Rosenkrantz, D.J., Stearns, R.E., Lewis II., P.M.: An analysis of several heuristics for the traveling salesman problem. SIAM Journal on Computing 6(3), 563–581 (1977)
20. Talbi, E.-G.: A taxonomy of hybrid metaheuristics. Journal of Heuristics 8(2), 541–564 (2002)
21. Zitzler, E., Künzli, S.: Indicator-based selection in multiobjective search. In: Yao, X., Burke, E.K., Lozano, J.A., Smith, J., Merelo-Guervós, J.J., Bullinaria, J.A., Rowe, J.E., Tiňo, P., Kabán, A., Schwefel, H.-P. (eds.) PPSN 2004. LNCS, vol. 3242, pp. 832–842. Springer, Heidelberg (2004)
22. Zitzler, E., Thiele, L.: Multiobjective evolutionary algorithms: A comparative case study and the strength pareto approach. IEEE Transactions on Evolutionary Computation 3(4), 257–271 (1999)
23. Zitzler, E., Thiele, L., Laumanns, M., Fonesca, C.M., Grunert da Fonseca, V.: Performance assessment of multiobjective optimizers: An analysis and review. IEEE Transactions on Evolutionary Computation 7(2), 117–132 (2003)

Boosting VNS with Neighborhood Heuristics for Solving Constraint Optimization Problems

Nicolas Levasseur, Patrice Boizumault, and Samir Loudni

GREYC, UMR60-72
Campus Côte de Nacre, boulevard du Maréchal Juin, BP 5186
14032 CAEN Cedex, France

Abstract. Weighted CSPs (Constraint Satisfaction Problems) are used to model and to solve constraint optimization problems using tree search or local search methods that use large neighborhoods. For the last ones, selecting the neighborhood to explore is crucial. Some heuristics defined for CSP, as ConflictVar, are based on conflicts. In this article, we propose new neighborhood heuristics for WCSP, not only based on conflicts, but also depending on the topology of the constraints graph and violation costs of constraints. Experiments performed with VNS/LDS+CP, a particular instance of VNS (Variable Neighborhood Search) we developed, show that our heuristics clearly outperform ConflictVar.

1   Introduction

Weighted Constraint Satisfaction Problems formalism [6,14] is a generic framework used to model and to solve constraint optimization problems which allows to deal with over-constrained problems and preferences between solutions. WCSP are solved by local, hybrid or tree search methods. Hybrid methods, by combining advantages of two others, provide a good trade-off between computing times and quality of solutions.

For local search methods that use large neighborhoods, the design of a neighborhood heuristic is crucial, since it chooses parts of the search space to explore in order to find solutions of better quality. However, for an effective tuning, defining crafted neighborhoods requires deep specific knowledge of the problem, as well as a lot of experiments. To our knowledge, the few heuristics defined for CSP, as ConflictVar, are based on conflicts (a variable is conflicted if it is related to at least one violated constraint).

In this paper, we propose new neighborhood heuristics dedicated to WCSP, not only based on conflicts, but also exploiting the topology of the constraint graph and violation costs of constraints. Experiments have been performed on real life instances (CELAR) and random instances (GRAPH) using VNS/LDS+CP (a particular instance of VNS [11] we developed for solving anytime problems [9,10]). These experiments show that our heuristics clearly outperform ConflictVar.

First, we present WCSP formalism (Section 2), then we give an overview of VNS/LDS+CP and recall neighborhood heuristics for CSP (Section 3). New

M.J. Blesa et al. (Eds.): HM 2008, LNCS 5296, pp. 131–145, 2008.
© Springer-Verlag Berlin Heidelberg 2008

neighborhood heuristics are described in Section 4, and experiments in Section 5. Finally, we conclude.

2   Weighted Constraint Satisfaction Problems

A Weighted Constraint Satisfaction Problem (WCSP) is defined by a quadruplet $(\mathcal{X}, \mathcal{D}, \mathcal{C}, k)$, with $\mathcal{X} = \{x_1, ..., x_n\}$ the set of variables (of size n), $\mathcal{D} = \{d_1, ..., d_n\}$ their finite domains (the maximal domain size is noted d) and $\mathcal{S}(k_\top)$ its valuation structure. $\mathcal{S}(k_\top)$ is a triplet $([0,1,...,k_\top], \oplus, \geq)$ where: $k_\top$ is a natural number in $[1, .., \infty]$, $\oplus$ is defined by: $a \oplus b = min(k_\top, a+b)$ and $\geq$ is the standard order operator among naturals. $\mathcal{C}$ is the set (of size e) of constraints. Each constraint $c \in \mathcal{C}$, is defined on a subset $\mathcal{X}_c \subseteq \mathcal{X}$ of related variables. $\mathcal{A}_{\downarrow \mathcal{X}_c}$ is the set of assignments for these variables. Each $c \in \mathcal{C}$ is defined by a function $f_c : \prod_{x_i \in \mathcal{X}_c} d_i \mapsto [0, k_\top]$. For a complete assignment $\mathcal{A}$, if $f_c(\mathcal{A}_{\mathcal{X}_c}) = 0$, c is said to be satisfied, else to be violated. An assignment of x_i to a value a is noted: $(x_i = a)$. A *complete* assignment (or solution) $\mathcal{A} = (a_1, ..., a_n)$ is an assignment of all variables; on the contrary, it will be called a *partial* assignment. The cost of a complete assignment $\mathcal{A} = (a_1, ..., a_n)$ is noted: $\mathcal{V}(\mathcal{A}) = \sum_{c \in \mathcal{C}} f_c(\mathcal{A}_{\downarrow \mathcal{X}_c})$. The objective is to find a complete assignment of minimal cost: $min_{\mathcal{A} \in d_1 \times d_2 \times ... \times d_n} \mathcal{V}(\mathcal{A})$.

3   VNS/LDS+CP

VNS/LDS+CP [8,9] is a local search method based on a Variable Neighborhood Search method with approximated decompositions (VNDS) [2]. Neighborhoods are obtained by unfixing a part of the current solution according to a neighborhood heuristic. Then the exploration of the search space related to the unfixed part of the current solution is performed by a partial tree search (LDS, [3]) with constraint propagation (CP). VNS/LDS+CP has been developed for solving anytime problems [7,9] and successfully applied to on-line resources allocation for ATM (Asynchronous Transfer Mode) networks with rerouting [10].

3.1   Principles

Algorithm 1 shows the pseudo-code of VNS/LDS+CP. It starts from an initial solution s which is randomly generated. A subset of k variables (k is the *dimension* of the neighborhood) is selected by the neighborhood heuristic Hneighborhood in N_k (a set of all subsets of k variables among $\mathcal{X}$) (line 5). A partial assignment $\mathcal{A}$ is generated from the current solution s by unassigning the k selected variables; the $(n - k)$ non-selected variables keep their current value in s (line 6). Then, unassigned variables are rebuilt by a partial tree search, LDS+CP combined with constraint propagation based on computation of lower bounds. If LDS+CP finds a solution of better quality s' in the neighborhood of s (line 8), then s' becomes the current solution and k is reset to k_{init} (lines 9-10). Otherwise, we look for improvements in the subspace where $(k + 1)$ variables will be unassigned (line 11). Indeed, higher is the dimension of the neighborhood,

Algorithm 1. VNS/LDS+CP

```
function VNS(𝒳, 𝒞, k_init, k_max, δ_max) begin
1    s ← genRandomSol(𝒳)
2    k ← k_init
3    iter ← 1
4    while (k < k_max) ∧ (notTimeOut) do
5        𝒳_unassigned ← Hneighborhood(N_k, s)
6        𝒜 ← s\{(x_i = a)s.t.x_i ∈ 𝒳_unassigned}
7        s' ← NaryLDS+CP(𝒜, 𝒳_unassigned, δ_max, 𝒱(s), s)
8        if 𝒱(s') < 𝒱(s) then
9            s ← s'
10           k ← k_init
11       else k ← k + 1
12   return s
end
```

larger is the search space and likely to contain better solutions than the current one. However, since the size of neighborhoods can quickly grow, finding the best neighbor may require a too expensive effort. That is why, in order to efficiently explore parts of the search space, we use LDS+CP, a partial search combined with constraint propagation. The algorithm stops when it reaches the maximal dimension size allowed or the timeout (line 4).

The efficiency of VNS/LDS+CP strongly depends on the strategy used to manage the neighborhood size, the heuristic which selects variables to unassign and the method used to rebuild them. In [9], authors have shown the importance of using a VNS strategy compared to a LNS one [15], and the relevance of rebuilding unassigned variables with LDS+CP.

3.2 ConflictVar

Neighborhood heuristics are crucial, since they select parts of the search space to explore in order to find solutions of better quality. But as quoted in the introduction, designing efficient heuristics is difficult and problem-dependent. To our knowledge, very few heuristics which are not driven by specific knowledge exist. ConflictVar, defined for CSP, is the most popular one. It is based on conflicted variables; for a complete assignment $\mathcal{A}$, a variable is said to be conflicted if it occurs in at least one violated constraint c $(f_c(\mathcal{A}_{\mathcal{X}_c}) \neq 0)$. Another heuristic, called PGLNS [13], proposed for optimization problems, uses effects of propagations to unassign variables that are inter-related.

For a given dimension of neighborhood k, first, ConflictVar randomly selects k variables to unassign among conflicted variables $(\mathcal{X}_{conflicted})$, then among all non-conflicted variables (if $k > \#(\mathcal{X}_{conflicted})$). Such a heuristic which is mainly based on random choices, allows to diversify the search and to quickly escape from local minima. The pseudo-code of ConflictVar is depicted in Algorithm 2; the function getConflict returns for an assignment $\mathcal{A}$, all conflicted variables.

Algorithm 2. ConflictVar

```
function ConflictVar (A, X, C, k) begin
```
$\quad$ $\mathcal{X}_{unassigned} \leftarrow \emptyset$

1

2 $\quad$ $\mathcal{X}_{conflicted} \leftarrow$ getConflict$(\mathcal{A}, \mathcal{C})$

3 $\quad$ **while** $\#(\mathcal{X}_{unassigned}) \neq k$ **do**

4 $\quad\quad$ **if** $\mathcal{X}_{conflicted} \neq \emptyset$ **then**

5 $\quad\quad\quad$ $x \leftarrow randomPick(\mathcal{X}_{conflicted})$

6 $\quad\quad\quad$ $\mathcal{X}_{conflicted} \leftarrow \mathcal{X}_{conflicted} \backslash \{x\}$

7 $\quad\quad$ **else** $x \leftarrow randomPick(\mathcal{X} \backslash \mathcal{X}_{unassigned})$

8 $\quad\quad$ $\mathcal{X}_{unassigned} \leftarrow \mathcal{X}_{unassigned} \cup \{x\}$

9 $\quad$ **return** $\mathcal{X}_{unassigned}$

```
end
```

ConflictVar is a simple heuristic and easy to implement since it is problem independent. However, this heuristic has two main drawbacks : it neither depends on the topology of the constraint graph and neither takes into account violation costs of constraints. For example, ConflictVar may only select unrelated variables (i.e. there is no constraint which is fully unassigned), and all selected variables may also have high degree (i.e. they occur in many constraints). In such a case, it is unlikely to rebuild them without violating many constraints, and thus to find a better solution.

3.3   Limited Discrepancy Search

LDS (Limited Discrepancy Search) is a tree search method introduced by Harvey and Ginsberg [3] allowing to iteratively solve binary CSP. Let H be a heuristic that is trusted. The main idea of LDS is to follow H when exploring the search tree, and to consider that H may mistake a small number (δ) of times. Thus, δ *discrepancies* are allowed during search. For a given maximal number δ_{max} of discrepancies, LDS explores the tree in an iterative way with an increasing number of *discrepancies* (from $\delta = 0$ to $\delta = \delta_{max}$). Depending on the value of δ_{max}, LDS is either a partial or a complete tree search. In [9], LDS have been extended to n-ary optimization problems, and performs only the last iteration (for $\delta = \delta_{max}$). Algorithm 3 shows pseudo-code of n-ary LDS+CP where $\mathcal{A}$ is a partial assignment, $\mathcal{X}_f$ is the set of future variables (i.e. not assigned yet), and δ the number of discrepancies allowed.

If all variables are assigned (line 1), the best known solution is updated (line 2) and the current solution becomes the best one (line 3). Otherwise, a future variable (i.e not assigned yet) $x_i \in \mathcal{X}_f$ is selected (line 4); let d_i be the current domain of x_i ordered by a heuristic H. In the while loop, a discrepancy equal to j corresponds to the value of rank $(j+1)$. As long as there are enough discrepancies ($j \leq \delta$) and enough values in d_i ($d_i \neq \emptyset$), the value a of rank $(j+1)$ is selected (line 7) and the remaining amount of discrepancies is set to $(\delta - j)$. Then constraint propagation (see Section 3.4) is performed on values of future variables in order

---

Algorithm 3. N-ary optimization LDS+CP

---

function NaryLDS+CP $(\mathcal{A}, \mathcal{X}_f, \delta, \mathcal{UB}, bestS)$ **begin**

1   **if** $\mathcal{X}_f = \emptyset$ **then**
2      $\mathcal{UB} \leftarrow \mathcal{V}(\mathcal{A})$
3   **return** $\mathcal{A}$

4    $x_i \leftarrow$ select-variable $(\mathcal{X}_f)$
5    $j \leftarrow 0$
6   **while** $(d_i \neq \emptyset) \wedge (j \leq \delta)$ **do**
7      $a \leftarrow$ select-value (d_i)
8   **if** $CP(\mathcal{A} \cup \{(x_i = a)\}, \mathcal{X}_f \setminus \{x_i\}, \mathcal{UB})$ **then**
9   bestS $\leftarrow$ NaryLDS+CP $(\mathcal{A} \cup \{(x_i = a)\}, \mathcal{X}_f \setminus \{x_i\}, \delta - j, \mathcal{UB}, bestS)$
10   undoCP $(\mathcal{A}, (x_i = a), \mathcal{X}_f)$
11      $d_i \leftarrow d_i \setminus \{a\},$
12      $j \leftarrow j + 1$

13   **return** bestS
end

---

to prune values that can not extend $\mathcal{A}$ to a complete assignment with a valuation better than the best known one. If there is no empty domain (line 8), search continues (line 9). Otherwise (or when the exploration of the sub-space is over), a backtrack is performed in order to undo all modifications performed during the constraint propagation (line 10). Then the domain of variable x_i is reduced (line 11), and x_i is assigned to the next value of its reduced domain (if it is not an empty one) (line 6).

3.4 Constraint Propagation

The aim of Constraint Propagation is, given a partial assignment $\mathcal{A}$, to remove values of future variables (i.e. not assigned yet) that could not occur in a solution with a better cost than the best known one. For this, at each node of the search tree, for each value a for each future variable x_i, a lower bound $\mathcal{LB}(\mathcal{A} \cup \{(x_i = a)\})$ representing the aggregation of costs of constraints which will be necessarily violated by any extension of $(\mathcal{A} \cup \{(x_i = a)\})$ to a complete assignment is computed. If $\mathcal{LB}(\mathcal{A} \cup \{(x_i = a)\})$ is greater than the cost of the best known solution, a is removed (i.e. the subtree is pruned). Several methods (*forward checking* and *directed arc-consistency* [5]) have been proposed to compute lower bounds for binary constraints.

The function $CP(\mathcal{A} \cup \{(x_i = a)\}, \mathcal{X}_f \setminus \{x_i\}, \mathcal{UB})$, used in Algorithm 3, establishes a local consistency at each node, and returns **false** if all values of a domain are removed, **true** if not. According to the consistency level wished, filtering can be carried out by various algorithms: Partial Forward Checking-Directed Arc Consistency (PFC-DAC), Maintaining Reversible DAC (PFC-MRDAC) [5], or Maintaining Full Directional Arc Consistency (MFDAC*) [6]. MFDAC* detects more inconsistencies than PFC-(MR)DAC (see [6] for more details).

4   Neighborhood Heuristics

In this section, we propose new neighborhood heuristics based on `ConflictVar` and dedicated to solve constraint optimization problems modelled as WCSP. Our heuristics make use of the topology of the constraint graph as well as violation costs of constraints. First, we introduce a few definitions.

Definition 1. *Neighboring variables, let x and y be two variables, x is a neighbor of y if and only if x and y occur in a constraint. The set of all neighboring variables of x is denoted $neighbors(x)$.*

Definition 2. *Degree of a variable, let x be a variable, the degree of x, noted $deg(x)$ is equal to the cardinality of $neighbors(x)$. For a complete assignment $\mathcal{A}$, the number of conflicted variables in the neighborhood of x will be noted $conflictDeg(x)$.*

Definition 3. *Degree of freedom of a variable, let $\mathcal{A}$ be a complete assignment, $\mathcal{X}_{unassigned}$ be a set of variables to unassign and x be a variable in $\mathcal{X}_{unassigned}$. The degree of freedom of x is equal to the number of variables in $neighbors(x)$ belonging to $\mathcal{X}_{unassigned}$. If $neighbors(x) \subset X_{unassigned}$, the freedom degree of x is said to be maximal.*

4.1   Extensions of ConflictVar

`ConflictVar` randomly selects a subset of conflicted variables independently of the topology of the constraint graph. Each selected variable may have a degree of freedom equal to zero. In this case, it is less likely that the rebuilding step finds a new solution of better quality by rebuilding the unassigned part of the current solution. In our experiments, we have noticed that, during the rebuilding step, higher is the degree of freedom of a variable, more opportunities the rebuilding step has to minimize inconsistencies. Thus, we propose to exploit the topology of the constraint graph to define new neighborhood heuristics which maximize degrees of freedom of selected variables.

Figures 1 shows some choices performed by our heuristics. Dashed lines correspond to constraints satisfied by the current complete assignment $\mathcal{A}$; others constraints are violated. The first variable selected by the neighborhood heuristics is in black; next selected ones are noted in gray. For each heuristic, the next selected variable, will be chosen among variables in dashed.

1. ConflictVar-Connected randomly selects the next variable among conflicted ones neighboring to those already selected (see figure 1a). The first variable is randomly selected among conflicted ones.

2. ConflictVar-Star selects a first variable x_c (in black) noted "center" (see figure 1b), then it randomly selects variables among conflicted ones included in $neighbors(x_c)$ (i.e. in gray). If all of them have been already selected, a new "center" variable is generated among conflicted ones neighboring to already selected variables (i.e. variables in dashed).

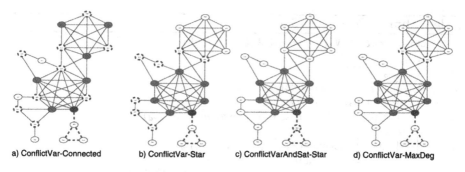

a) ConflictVar-Connected   b) ConflictVar-Star   c) ConflictVarAndSat-Star   d) ConflictVar-MaxDeg

Fig. 1. Neighborhood heuristics

Algorithm 4. ConflictVar-Connected

```
function ConflictVar-Connected (𝒜, 𝒳, 𝒞, k) begin
1    𝒳unassigned ← ∅
2    𝒳allowed ← ∅
3    𝒳conflicted ← getConflict(𝒜, 𝒞)
4    while #(𝒳unassigned) ≠ k do
6        if 𝒳allowed ≠ ∅ then  x ← randomPick(𝒳allowed)
8        else if 𝒳conflicted ≠ ∅ then  x ← randomPick(𝒳conflicted)
9        else x ← randomPick(𝒳\𝒳unassigned)
10       𝒳unassigned ← 𝒳unassigned ∪ {x}
11       𝒳conflicted ← 𝒳conflicted\{x}
12       𝒳allowed ← (𝒳allowed\{x}) ∪ (neighbors(x) ∩ 𝒳conflicted)
13   return 𝒳unassigned
   end
```

---

3. ConflictAndSatVar-Star, which extends **ConflictVar-Star**, enlarges when all conflicted variables neighboring to the "center" variable have been selected, the choice of variables to those among the non-conflicted ones neighboring to the center. In figure 1c, the next variable to be chosen is the one in dashed.

4. ConflictVar-MaxDeg first, selects a variable among conflicted ones. Then, next variables are chosen among ones (conflicted or non-conflicted) having a maximal number of neighbors already selected (ties are randomly broken). In figure 1d, each variable in dashed has two selected neighboring variables, whereas other variables have only one.

5. ConflictVar-H-Cost uses violation costs of constraints to select variables to unassign. See Section 4.6 for more details.

4.2   ConflictVar-Connected

The ConflictVar-Connected heuristic allows to increase to 1 minimal degrees of freedom of selected variables. Its pseudo-code is detailed by Algorithm 4.

ConflictVar-Connected selects k variables to unassign. Each variable is randomly selected among the first following nonempty subset:

Algorithm 5. ConflictVar-Star

---

function ConflictVar-Star $(\mathcal{A}, \mathcal{X}, \mathcal{C}, k)$ **begin**

1     $\mathcal{X}_{unassigned} \leftarrow \emptyset$
2     $\mathcal{X}_{allowed} \leftarrow \emptyset$
3     $\mathcal{X}_{conflicted} \leftarrow \texttt{getConflict}(\mathcal{A}, \mathcal{C})$
4   **while** $\#(\mathcal{X}_{unassigned}) \neq k$ **do**
6   **if** $\mathcal{X}_{allowed} \neq \emptyset$ **then** $x \leftarrow randomPick(\mathcal{X}_{allowed})$
       else
7           $x \leftarrow \texttt{GetCenter}(\mathcal{X}_{unassigned}, \mathcal{X}_{conflicted})$
8   **if** $x = null$ **then**
9   **if** $\mathcal{X}_{conflicted} \neq \emptyset$ **then** $x \leftarrow randomPick(\mathcal{X}_{conflicted})$
10   **else** $x \leftarrow randomPick(\mathcal{X} \backslash \mathcal{X}_{unassigned})$
11           $\mathcal{X}_{allowed} \leftarrow (neighbors(x) \cap \mathcal{X}_{conflicted})$
12        $\mathcal{X}_{unassigned} \leftarrow \mathcal{X}_{unassigned} \cup \{x\}$
13        $\mathcal{X}_{conflicted} \leftarrow \mathcal{X}_{conflicted} \backslash \{x\}$
14        $\mathcal{X}_{allowed} \leftarrow \mathcal{X}_{allowed} \backslash \{x\}$
15   **return** $\mathcal{X}_{unassigned}$
end

function GetCenter $(\mathcal{X}_{unassigned}, \mathcal{X}')$
begin
16     $\mathcal{X}'' \leftarrow \cup_{x \in \mathcal{X}_{unassigned}}(neighbors(x) \cap \mathcal{X}')$
17   **if** $\mathcal{X}'' \neq \emptyset$ **then return** $randomPick(\mathcal{X}'')$
18   **return** $null$
end

---

1. the subset $\mathcal{X}_{allowed}$ of conflicted variables neighboring to already selected ones (lines 5 and 6),
2. the subset of conflicted variables (lines 7 and 8),
3. the subset of not yet selected variables ($\mathcal{X} \backslash \mathcal{X}_{unassigned}$) (line 9).
When a variable x is selected, it is added to $\mathcal{X}_{unassigned}$ (line 10), removed (if required) from conflicted variables (line 11), and all conflicted variables in $neighbors(x)$ are allowed to be selected later (i.e. added to $\mathcal{X}_{allowed}$) (line 12).

The degree of freedom of each selected variable is between a minimal value equal to 1 (since all variables are connected) and a maximal value equal to its degree. However, in our experiments, we have noticed that this maximal value is never reached. Indeed, since the size of $\mathcal{X}_{allowed}$ grows quickly, probability for a variable x and all its neighbors to be selected significantly decreases.

4.3   ConflictVar-Star

The heuristic ConflictVar-Star, which improves ConflictVar-Connected, tries to maximize the degree of freedom of at least one selected variable. Its pseudo-code is detailed by Algorithm 5.

The function GetCenter randomly selects a "center" variable among conflicted variables neighboring to selected ones.

First, the heuristic `ConflictVar-Star` selects a center variable x_c, then randomly chooses next variables among conflicted ones in $neighbors(x_c)$. If all of these are selected, a new center is chosen in the first following nonempty subset:
1. the subset of conflicted variables neighboring to selected ones (line 7),
2. the subset of conflicted variables among $\mathcal{X}_{conflicted}$ (line 9),
3. the subset of not selected variables yet (line 10).
Then, neighbors of the new center are allowed to be selected (line 11). As before, when a variable is selected, $\mathcal{X}_{unassigned}$, $\mathcal{X}_{conflicted}$ and $\mathcal{X}_{allowed}$ are updated (lines 12-14).

The degree of freedom of the first center x_{c1} is equal to : $max(k-1, conflictDeg(x_{c1}))$. Degrees of freedom of other variables are still between 1 and their degree.

4.4   ConflictVarAndSat-Star

In this section, we show the relevance of selecting some non-conflicted variables. Let $\mathcal{A}$ be a complete assignment with $(x_1 = a)$ and $(x_2 = a)$, and suppose, for the problem in Figure 1, that the constraint c in dashed related to x_1 (in black) and to x_2 is a hard equality constraint. We remind that constraints in dashed lines are satisfied by $\mathcal{A}$. A heuristic, only based on conflicts, would select x_1, but not x_2, since it is not a conflicted variable. In this case, during the rebuilding step, x_1 will never be reassigned to another value (since c is a hard constraint). So, we propose to extend `ConflictVarStar` to `ConflictVarAndSat-Star` which also selects non-conflicted variables.

As before, first, `ConflictAndSatVarStar` randomly selects a center variable x_c among $\mathcal{X}_{unassigned}$, then allows conflicted variables among $neighbors(x_c)$ to be chosen. If all of these variables are already selected, before choosing a new center variables, next variables will be selected among non-conflicted ones in $neighbors(x_c)$. Each time, a new center has to be chosen, it is selected among conflicted variables neighboring to selected ones (if some are not selected yet), otherwise among non-conflicted variables neighboring to selected ones.

The degree of freedom of the first center is equal to : $max(k - 1, deg(x_{c1}))$, with $deg(x_{c1}) \geq conflictDeg(x_{c1})$. Degrees of freedom of other variables are still between 1 and their degree.

4.5   ConflictVar-MaxDeg

We also propose the heuristic `ConflictVar-MaxDeg`, which tries to maximize degrees of freedom of all selected variables. Algorithm 6 details its pseudo-code.

The function `GetVarsMaxCard` randomly selects a (conflicted or non-conflicted) variable among those having the highest number of neighboring variables already selected (ties are randomly broken).

First, `ConflictVar-MaxDeg` selects a variable x among conflicted ones. Then, for each variable x' in $neighbors(x)$, $card[x']$ (i.e. the number of already selected neighbors of x') is increased by 1. Next variables will be randomly chosen among (conflicted or non-conflicted) variables with the highest value of $card$.

The degree of freedom may still be equal to 1 in the worst case. However, in practice we have noticed that, degrees of freedom of selected variables are

---

Algorithm 6. ConflictVar-MaxDeg

```
    function ConflictVar-MaxDeg (A, X, C, k) begin
1       Let card be an array of size n
2       x ← randomPick(getConflict(A, C))
3       X_unassigned ← {x}
4       foreach x' ∈ neighbors(x) do  card[x'] ← card[x'] + 1
6       while #(X_unassigned) ≠ k do
7           x ← GetVarsMaxCard(X \ X_unassigned, card)
8           X_unassigned ← X_unassigned ∪ {x}
10          foreach x' ∈ neighbors(x) do  card[x'] ← card[x'] + 1
11      return X_unassigned
    end
```

---

higher. Nevertheless, this heuristic less depends on randomness than previous ones, since it is more guided by a deterministic criterion.

4.6   ConflictVar-H-Cost

In the WCSP framework, it is important to try to satisfy, as often as possible, constraints with high violation cost. However, some of them may be violated in all optimal solutions. So, heuristics defined for WCSP should try to satisfy constraints with a high violation cost (by regularly unassigning its variables), but without trying to satisfy them all the time.

To achieve this goal, we propose to extend each previous heuristic in order to make them dependent on violation costs as follows (the extension of a heuristic H is noted H-Cost). For a current complete assignment A, first, costs of all constraints are sorted in decreasing order, and divided into $nbSets$ subsets noted $ec_1, ec_2, ...ec_{nbSets}$, with $nbSets$ a parameter to set at the beginning of the search. Each subset ec_i contains the $(i * e)/nbSets$ highest violation costs (with e the number of constraints). During the search, only constraints with a violation cost higher or equal to $min(ec_b)$ will be considered as conflicted; the value of b proportionally increases according to the value of k. When k is equal to k_{min} (resp. k_{max}), b is equal to 1 (resp. $nbSets$). So, during the search (if the timeout is not reached), all constraints will be considered as conflicted at least once. Besides, variables involved in constraints with a high violation cost (for the current assignment), will be unassigned more regularly (since k is reset each time a new solution is found). For example, for 6 constraints and $nbSets$ equal to 2, violations costs of constraints are sorted $\{14, 2, 1, 1, 0, 0\}$ and divided into the two subsets $ec_1 = \{14, 2, 1\}$ and $ec_2 = \{14, 2, 1, 1, 0, 0\}$.

In this paper, we only detail the extension of ConflictVar (see Algorithm 7). However, the extension of ConflictVar-Star has also been considered.

In Algorithm 7, the function InitMinEc returns an array, noted $minec$, containing all $min(ec_i)$ previously defined. The function getB returns the value of b according to the current dimension of the neighborhood. The function getOptConflict returns a subset of variables considered as conflicted (i.e. involved in a constraint with a violation cost higher or equal to $minec[b]$).

Algorithm 7. ConflictVar-Cost

---

function ConflictVar-Cost $(\mathcal{A}, \mathcal{X}, \mathcal{C}, k, k_{init}, k_{max}, nbSets)$ **begin**

1   Let $minec[]$ be an array (of size $nbSets$) of integers

2     $minec \leftarrow$ InitMinEc$(\mathcal{A}, minec, \mathcal{C}, nbSets)$

3     $b \leftarrow$ getB$(k, k_{init}, k_{max}, nbSets)$

4     $\mathcal{X}_{unassigned} \leftarrow \emptyset$

5     $\mathcal{X}_{conflicted} \leftarrow$ getOptConflict$(\mathcal{A}, \mathcal{C}, minec[b])$

6   **while** $\#(\mathcal{X}_{unassigned}) \neq k$ **do**

7   **while** $\mathcal{X}_{conflicted} = \emptyset$ **do**

8           $b \leftarrow b + 1$

9           $\mathcal{X}_{conflicted} \leftarrow$ getOptConflict$(\mathcal{A}, \mathcal{C}, minec[b]) \setminus \mathcal{X}_{unassigned}$

10        $x \leftarrow randomPick(\mathcal{X}_{conflicted})$

11        $\mathcal{X}_{unassigned} \leftarrow \mathcal{X}_{unassigned} \cup \{x\}$

12        $\mathcal{X}_{conflicted} \leftarrow \mathcal{X}_{conflicted} \setminus \{x\}$

13   **return** $\mathcal{X}_{unassigned}$

end

function getB $(k, k_{init}, k_{max}, nbSets)$ **begin**

14   **return** $\lfloor 1 + (nbSets - 1) * \frac{k - k_{init}}{k_{max} - k_{init}} \rfloor$

end

ConflictVar-Cost randomly selects variables among conflicted ones (line 10). If all of them are chosen (line 7), the set of constraints considered as conflicted is enlarged to constraints with lower violation costs according to values in $minec$ (lines 8-9).

5   Experiments

RLFAP Instances: The CELAR (French: Centre d'Electronique de l'Armement) has made available a set of instances for the Radio Link Frequency Assignment Problem (RLFAP [1]). They consist in assigning a limited number of frequencies to a set of radio links defined between pairs of sites, in order to minimize interferences due to the re-use of frequencies. For experiments, we have considered instances from Scen06 to Scen10. A simplification is performed on each instance before search in order to satisfy all hard equality constraints; these constraints are, because of the nature of the domain of their related variables, bijective.

GRAPH instances: Generating Radio link frequency Assignment Problems Heuristically ([16]) is a random generator of instances similar to CELAR ones.

Each instance has been solved by VNS/LDS+CP, with a discrepancy of 4, which is the best value found on RLFAP instances (see [9]). k_{min}, k_{max} have been set to 5 and 25, and the timeout to 6 minutes. All search strategies have been implemented in Java using the library *choco* [4]. Each curve represents an average over 100 resolutions of the evolution of the best known solution. Experiments have been performed on a 1.6 GHz P4 processor. For ConflictVar-H-Cost heuristics, $nbSets$ has been arbitrarily set to 5.

Figures from 2 to 6 (resp. from 7 to 12) depict results obtained on RLFAP (resp. GRAPH) instances. Results allow us to make the following observations :

1. for all instances, performances of `ConflictVar` and `ConflictVar-Connected` are very similar. So, even if the minimal value of degree of freedom is increased by 1, this is not sufficient to identify assignments which need to be rebuilt.

2. Heuristics with a higher degree of freedom, are clearly more relevant. For all instances (except those where results are similar and Scen10), `ConflictVar-MaxDeg` outperforms `ConflictVar-Star` and `ConflictVarAndSat-Star`, in particular on Scen06 and Scen07 (which are among the most difficult ones).

3. Selecting non-conflicted variables is an important criterion. Indeed, `ConflictVarAndSat-Star` outperforms `ConflictVar-Star` (except on Scen10).

4. `ConflictVar-MaxDeg` which combines the two previous criteria (by maximizing the degree of freedom of selected variables, and choosing also non-conflicted ones), is a very efficient heuristic. Indeed, on most instances, it is one of the best ones (except on Scen10).

5. Compared to `ConflictVar`, `ConflictVar-Cost` performs better on most instances. Moreover, the combination of `ConflictVar-Star` with cost selection (i.e. `ConflictVar-Star-Cost`) outperforms `ConflictVar-MaxDeg` on Scen10 and presents the same behavior on GRAPH instances. This confirms that it could be important to consider violation costs of constraints when choosing variables to be unassigned.

Table 1 shows for each instance, the cost of the best solution found by each heuristic. Each heuristic is noted by a number, best results are in bold, and best solutions which are not proved optimal are in italic. `ConflictVar-Star-Cost` and `ConflictVar-MaxDeg` are the best heuristics, with a small advantage to the first one which has found optimal solutions on 4 GRAPH instances, 2 CELAR instances, and is really very close to the optimum on other instances. Our results are quite similar to those obtained by `ID Walk` [12] on CELAR instances.

Table 1. Best solutions found on each instance

	s-06	s-07	s-08	s-09	s-10	g-05	g-06	g-07	g-11	g-12	g-13
Optimum	3389	*343592*	*262*	15571	31516	221	4123	4324	*3080*	11827	*10110*
ConflictVar	3434	2599275	490	15675	31518	332	12434	4368	25183	11846	38149
CV-Connected	3514	2394230	491	15821	**31516**	320	10563	4383	22911	11839	37944
CV-MaxDeg	3401	354302	**286**	**15571**	**31516**	**221**	**4123**	**4324**	3350	**11827**	**11599**
CV-Star	3528	364410	343	**15571**	**31516**	290	4250	**4324**	5380	**11827**	15384
CVAndSat-Star	3412	364103	**286**	**15571**	**31516**	**221**	4126	**4324**	3473	**11827**	12335
CV-Cost	3463	586159	468	**15571**	**31516**	250	8277	**4324**	19112	11828	35606
CV-Star-Cost	**3399**	**343800**	296	**15571**	**31516**	**221**	**4123**	**4324**	3235	**11827**	12794

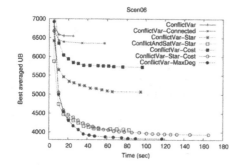

Fig. 2. Scen06 (n=100,e=1222)

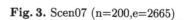
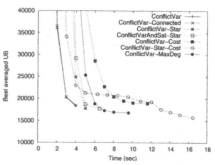

Fig. 3. Scen07 (n=200,e=2665)

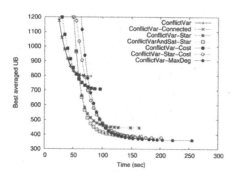

Fig. 4. Scen08 (n=458,e=5286)

Fig. 5. Scen09 (n=200,e=4209)

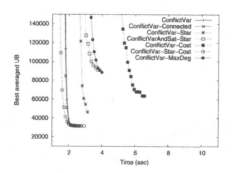

Fig. 6. Scen10 (n=200,e=4209)

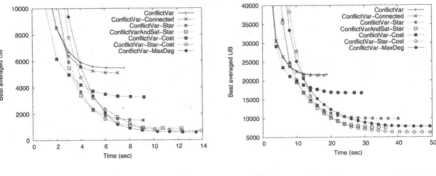

Fig. 7. Graph05 (n=100,e=1034)   **Fig. 8.** Graph06 (n=200,e=1970)

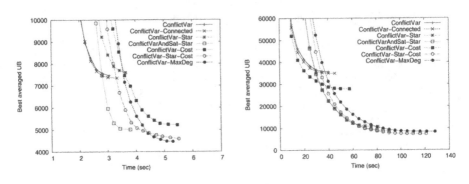

Fig. 9. Graph07 (n=141,e=2213)   **Fig. 10.** Graph11 (n=340,e=3417)

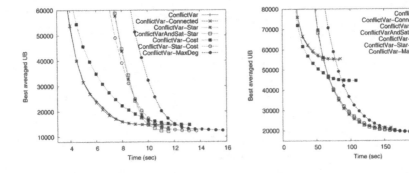

Fig. 11. Graph12 (n=252,e=4099)   **Fig. 12.** Graph13 (n=458,e=4815)

6   Conclusion

In this article, we have proposed new neighborhood heuristics dedicated to solve optimization problems modelled as WCSP based on conflict variables, graph topology and violation cost of constraints. Experiments performed with VNS/LDS+CP show that our heuristics clearly outperform ConflictVar.

In future works, we will study the influence of the value of the *nbSets* parameter on the efficiency of our cost-based heuristic. We also plan to compare our heuristics to PGLNS on RLFAP and random instances.

References

1. Cabon, B., de Givry, S., Lobjois, L., Schiex, T., Warners, J.P.: Radio link frequency assignment. Constraints 4(1), 79–89 (1999)
2. Hansen, P., Mladenovic, N., Perez-Brito, D.: Variable neighborhood decomposition search. Journal of Heuristics 7(4), 335–350 (2001)
3. Harvey, W.D., Ginsberg, M.L.: Limited discrepancy search. In: Proceedings of IJCAI 1995, pp. 607–615 (1995)
4. Jussien, N., Barichard, V.: The PALM system: explanation-based constraint programming. In: Proceedings of TRICS, workshop of CP, pp. 118–133 (2000)
5. Larrosa, J., Meseguer, P., Schiex, T.: Maintaining reversible DAC for solving MAX-CSP. Artificial Intelligence 107(1), 149–163 (1999)
6. Larrosa, J., Schiex, T.: In the quest of the best form of local consistency for Weighted CSP. In: Proceedings of IJCAI 2003, pp. 239–244 (2003)
7. Loudni, S., Boizumault, P.: VNS/LDS+CP: a hybrid method for constraint optimization in anytime contexts. In: Proceedings of MIC 2001, Porto, Portugal, pp. 761–765 (July 2001)
8. Loudni, S., Boizumault, P.: Combining VNS with constraint programming for solving anytime optimization problems. EJOR 191, 705–735 (2008)
9. Loudni, S., Boizumault, P.: Solving constraint optimization problems in anytime contexts. In: Proceedings of IJCAI 2003, pp. 251–256 (2003)
10. Loudni, S., Boizumault, P., David, P.: On-line resources allocation for ATM networks with rerouting. Computers & Operations Research 33(11), 2891–2917 (2006)
11. Mladenović, N., Hansen, P.: Variable neighborhood search. Computers & Operations Research 24, 1097–1100 (1997)
12. Neveu, B., Trombettoni, G., Glover, F.: ID Walk: A candidate list strategy with a simple diversification device. In: Wallace, M. (ed.) CP 2004. LNCS, vol. 3258, pp. 423–437. Springer, Heidelberg (2004)
13. Perron, L., Shaw, P., Furnon, V.: Propagation guided large neighborhood search. In: Wallace, M. (ed.) CP 2004. LNCS, vol. 3258, pp. 468–481. Springer, Heidelberg (2004)
14. Schiex, T., Fargier, H., Verfaillie, G.: Valued constraint satisfaction problems: Hard and easy problems. In: IJCAI 1995, pp. 631–639 (1995)
15. Shaw, P.: Using constraint programming and local search methods to solve vehicle routing problems. In: Maher, M.J., Puget, J.-F. (eds.) CP 1998. LNCS, vol. 1520, pp. 417–431. Springer, Heidelberg (1998)
16. van Benthem, H.: Graph: Generating radiolink frequency assignment problems heuristically (1995)

Hybrid Local Search Techniques for the Generalized Balanced Academic Curriculum Problem

Luca Di Gaspero and Andrea Schaerf

DIEGM, University of Udine
via delle Scienze 208, I-33100, Udine, Italy
l.digaspero@uniud.it, schaerf@uniud.it

Abstract. The Balanced Academic Curriculum Problem (BACP) consists in assigning courses to teaching periods satisfying prerequisites and balancing students' load. BACP is included in CSPlib along with three benchmark instances. However, the BACP formulation in CSPlib is actually simpler than the real problem that, in general, universities have to solve in practice.

In this paper, we propose a generalized formulation of the problem and we study a set of hybrid solution techniques based on high-level control strategies that drive a collection of basic local search components. The result of the study allows us to build a complex combination of simulated annealing, dynamic tabu search and large-neighborhood search. In addition, we present six new instances obtained from our university, which are much larger and more challenging than the CSPlib ones (the latter are always solved to optimality in less than 0.1 seconds by our techniques).

For the sake of possible future comparisons, we make available through the web all the input data, our scores and results, and a solution validator.

1   Introduction

The Balanced Academic Curriculum Problem (BACP) is an assignment problem that arises in universities, and consists in assigning courses to teaching periods satisfying prerequisites and balancing students' load.

A formulation of BACP has been proposed by Castro and Manzano [3], and it has been included in CSPLib [7, prob. 30] along with three benchmark instances.

The BAC problem, in the CSPLib formulation, has been also tackled by Hnich et al [9] and Castro et al [2], using CP and IP techniques, and by Lambert et al [11], using a hybrid techniques composed by genetic algorithms and constraint propagation. In all works, the authors report the finding of a proven-optimal solution for all three instances, although with quite different running times that range from less than one second to hundreds of seconds.

The BACP formulation is actually simpler than the real problem that universities have to solve in practice, at least for the cases we are aware of. In this

M.J. Blesa et al. (Eds.): HM 2008, LNCS 5296, pp. 146–157, 2008.
© Springer-Verlag Berlin Heidelberg 2008

paper, we try to overcome this limitation, and we define a more complex formulation, which applies, among others, to our institution (School of Engineering of University of Udine).

We study the application of general high-level local search strategies and we develop a hybrid solution technique based on a complex combination of simulated annealing, dynamic tabu search and large-neighborhood search. We propose six new instances obtained from our university. Our instances are much larger and (for our techniques) more challenging than the CSPLib ones, and their optimal value is not known. Furthermore, they show different structures, as they represent cases very different from each other.

We report our experimental analysis for these new instances. In addition, given that the formulation proposed here is actually a generalization of the BACP one, we have been able to test our solver also on the CSPLib instances. The outcome has been that all three CSPLib instances are solved to optimality in less than 0.1 seconds in almost all runs of our solvers. This proves that, on those simpler cases, the solvers' performances are comparable or better than state of the art solutions.

For the general problem and for the larger instances, obviously we have no "competitors" to compare to, besides the variants implemented by ourselves. Nevertheless, for the sake of possible future comparisons, we make available through the web (http://www.diegm.uniud.it/satt/projects/bacp/) all the input data, our results, and also the source code of the solution *validator*.

The validator is a simple program that takes two command-line parameters, an instance and a solution, and returns both a detailed list of the violations and a summary of the costs. As we discussed in detail in [13], we believe that the publication of the validator is indeed necessary when proposing a new problem, so as to prevent possible misunderstandings on the details of the formulation, and thus to provide against the publication of incorrect results.

2   Problem Formulations

We first present here the basic BACP formulation as proposed in [3]. Later we discuss the extensions that we have added for dealing with our real-world problem.

2.1   BACP Formulation

The basic formulation consists of the following entities and constraints:

Courses: Each course has an integer number of *credits*, and it has to be taught during the planning horizon of the university degree.
Periods: The planning horizon is composed by a given number of *teaching periods* that have to be assigned to courses. Periods are divided in years, and each year is divided in a fixed number of terms. For example, a 3-year degree organized in four trimesters per year has 12 periods.

Load limits: For each period there is a minimum and a maximum number of courses that can be assigned to it. Further, there are minimum and maximum limits also on the number of total credits per period.

Prerequisites: Based on their content, some courses have to be taught before other courses. This means that we are given a set of pairs of courses, such that the period assigned to the first course has to be strictly less than the period assigned to the second. Obviously, prerequisite relation is transitive and cannot contain cycles.

The problem consists in finding an assignment of courses to periods that satisfies all above (hard) constraints: load limits and prerequisites. The objective function accounts for the balancing of credits in periods. In detail, the objective function (to be minimized) used in the cited papers is: *the maximum number of total credits per period.*

For example, the CSPLib instance `bacp8` has 46 courses for a total of 133 credits and 8 periods. The average number of credits per period is $133/8 = 16.625$. Therefore, the lower bound of the maximum number of credits per period is 17. Solutions with value 17 are thus optimal.

2.2   GBACP Formulation

Given the above basic formulation as starting point, we extend it in the following directions.

Curricula: First of all, in the formulation it is implicitly assumed that a student takes all courses without personal choices, whereas in practice a student can select among alternatives. A *curriculum* is a set of courses representing a possible complete selection of a student. For each single curriculum, courses have to be balanced and limited in number. We extend the problem by considering many curricula, which may share some of the courses.

Preferences: Professors can express preferences about their teaching periods. Specifically, a teacher expresses his/her preferences for a specific term of the year but not for the year. A preference of a course for a given term results in a penalty for any assignment of the course to a period which is not in that term. Preferences are not strict, and therefore they contribute to the objective function (soft constraints).

The objective function is thus a composition of preference violations and un-balanced load. In order to adhere more precisely to the real situation, the load component that we consider is not based only on the maximum load, but it sums up all the deviations (positive and negative) from the average number of credits per period for each curriculum.

More precisely, for each curriculum we compute α as the total number of credits of a curriculum divided by the number of periods (not necessarily integer-valued). A number of credits per period equal to $\lfloor \alpha \rfloor$ or $\lceil \alpha \rceil$ has penalty 0. All values below $\lfloor \alpha \rfloor$ or above $\lceil \alpha \rceil$ are penalized. In order to avoid large discrepancies from α, the penalty is quadratic. That is, a deviation of 1, 2, or 3 counts as 1, 4, and 9 points of penalty respectively (and so on for larger values).

Load limits are evaluated for each single curriculum and summed up. Conversely, prerequisites remain expressed at the global level. In order to have a single objective function, we assign to each violation of a preference 5 points of penalty.

As mentioned above, the original formulation includes limits per period in terms of both number of courses and total credits. However, the quadratic penalty of credit balancing makes the presence of hard limits in the number of credits meaningless in our formulation, because large discrepancies are already prevented. Therefore, we remove this constraint. Conversely, the limits on the number of courses instead are maintained because they helps in avoiding extreme situations in terms of the number of final exams (independently of credits), which is desirable in our institution.

We call the corresponding problem GBACP (G for Generalized). Notice that a 0 cost solution in our GBACP formulation, costs $\lceil \alpha \rceil$ in the original BACP and it is thus optimal.

The opposite is not necessarily true. For example, for the CSPLib instance bacp8, which has (for its single curriculum) $\alpha = 16.625$, a solution with 15, 16 and 17 credits per period has optimal value (17) for the basic formulation, but the periods with 15 credits are penalized 1 in our formulation.

3   Local Search for GBACP

Local search is a family of general-purpose techniques for search and optimization problems, which are based on an iterative process of navigating a search space stepping from one solution to a neighboring one. The process is guided by a cost function, which drives the search toward good solutions.

In order to apply local search to GBACP we have to define several features. We first illustrate the search space and the procedure for computing the initial state. Then, we define the neighborhood structure, and finally we describe the set of search components employed and the high-level strategies for combining them.

3.1   Search Space, Initial Solution, and Neighborhood Relation

Based on prerequisites, we can infer some infeasible assignment by constraint propagation. For example, if course a must be before course b, then course a cannot be assigned to the last period, and course b cannot assigned to the first.

We therefore compute the transitive closure of the prerequisites, and based on it we determine a minimum and a maximum starting period for each course as its assignable range. The search space is then defined by all possible assignments of a period in the above range to each course.

Assignments that violates hard constraints (load limits and prerequisites) are also considered, and the violations (*distance to feasibility*) are included in the cost function of the search with a higher weight w.r.t. the objective function.

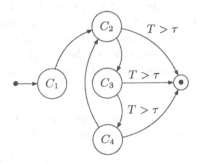

Fig. 1. An example of Generalized Local Search Machine with three search components

The initial solution is generated in a totally random way: each course is assigned a uniform random period in its assignable range.

The neighborhood relation consists in moving one course from its period to another one (in its range). A move is therefore identified by two attributes, namely the course and the new period.

3.2   Search Techniques

In this work we study a set of high-level search control strategies that hybridize several basic local search components. This idea is an instantiation of Hoos and Stützle's *Generalized Local Search Machines* (GLSM) [10, Chapter 3], which is a formal framework for describing search control by clearly separating it from the search components. In this framework, the basic search components are represented as states (i.e., nodes) of a Finite State Machine, whereas the transitions (i.e., edges) correspond to conditions for modeling the search control. Within the GLSM framework it is possible to specify also complex strategies such as VNS [8] and ILS [12].

In our terminology we refer to the states of the machine as *runners*, that are basic local search metaheuristics, or *kickers*, that are perturbation components represented by a single move in a large neighborhood used either for intensification or diversification purposes.

In Figure 1 we show an example of a GLSM with three search components. The search starts from C_1 and when this component has finished it unconditionally passes its solution to component C_2. Component C_2 continues the search, followed by C_3 and afterward C_4. Then the process is started again from C_2. The whole strategy is stopped when an overall timeout τ has expired.

3.3   Search Components for GBACP

The search components employed in this study range from very trivial strategies to more complex meta-heuristics. In detail, we consider the following ones (see [10] for a detailed description):

Steepest Descent (SD): at each step of the search the whole neighborhood is explored and the *best* improving move is performed. The search is stopped when a local minimum is reached.

First Descent (FD): differently from SD, this technique interrupts the exploration the neighborhood as soon as the *first* improving neighbor has been found. Like SD, The search is stopped when a local minimum is reached.

Randomized Hill Climbing (RHC): at each step of the search a random neighbor is selected. The move is performed only if the cost of the neighbor is less *or equal* than the current solution cost.

Simulated Annealing (SA): similarly to RHC a random neighbor is selected at each step. The move is performed either if it is an improving one or according to an exponential time-decreasing probability. In detail, if the cost of the move is $\Delta f > 0$, the move is accepted with probability $e^{-\Delta f/T}$, where T is a time-decreasing parameter called *temperature*. At each temperature level a number σ_N of neighbors of the current solution is sampled and the new solution is accepted according the above mentioned probability distribution. Afterward the value of T is modified using a *geometric* schedule, i.e., $T' = \beta \cdot T$, in which the parameter $\beta < 1$ is called the *cooling rate*.

Tabu Search (TS): at each step a subset of the neighborhood is explored and the neighbor that gives the minimum cost value becomes the new solution independently of the fact that its cost value is better or worse than the current one. The subset is induced by the *tabu list*, i.e., a list of the moves recently performed, which are currently forbidden and thus excluded from the exploration. Our tabu search implementation employs a dynamic short-term tabu list (called Robust Tabu Search in [10]), so that a move is kept in the tabu list for a random number of iterations in the range $[k_{min}..k_{max}]$.

Dynamic Tabu Search (DTS): the search strategy is the same as TS, however this variant of the algorithm is equipped with a mechanism that adaptively changes the shape of the cost function. In detail, the constraints which are satisfied for a given number of iterations will be relaxed (the weight is reduced by a factor $\gamma > 1$, i.e., $w' = w/\gamma$) in order to allow the exploration of solutions where those constraints do not hold. Conversely, if some constraint is not satisfied, it is tighten ($w' = w \cdot \gamma$) with the aim of driving the search toward its satisfaction.

Kickers (K): as explained in [6], kickers are special-purpose components that perform just one single step in a composite neighborhood (i.e., a sequence of moves of arbitrary length). Kickers support three strategies for selecting the moves: (i) *random kick* (K_r), that is a sequence of random moves, (ii) *first kick* (K_f), the first improving sequence in the exploration of the composite neighborhood, (iii) *best kick* (K_b), the best sequence in the exhaustive exploration of the composite neighborhood. Random kicks can be used for diversification purposes (thus giving raise to the Iterated Local Search strategy [12]), while first and best kicks are employed for intensifying the search.

3.4 GLSM Templates

The search components studied in this work are combined by means of the GLSM templates shown in Figure 2. These machines represent a set of fundamental strategies for the cooperation of search components.

The *multi-start* search (Fig. 2a) consists in the repetition of the generation of a starting solution, e.g., through a random assignment G_r, followed by the run of a basic meta-heuristic. The *multi-run* strategy (Fig. 2b), instead, repeats only the run of the meta-heuristic from the best solution found that far. These two strategies are simple strategies whose rationale is respectively to enhance diversification and intensification of a single runner at a coarse grain of granularity. The strategies are denoted in the following by $MS(R)$ and $MR(R)$, respectively.

The other two strategies, instead, include intensification/diversification during the high-level search process, according to the types of search components employed. The token-ring strategy (Fig. 2c) executes in sequence a set of basic meta-heuristic $R_1, R_2, \ldots$, followed by another search component that can either be a *kicker*, giving raise to what we call *run and kick* strategy (which is similar to Iterated Local Search [12]), or another *runner* (see [6] for more details). At the end of the process, the sequence is started again from R_1. The last template (Fig. 2d) is similar to the previous one, the only difference is that more emphasis is put on intensification, since the last component is a kicker that it can be iterated while it keeps improving the solution, thus performing a Large Neighborhood Search (LNS) [1]. Even though the difference between these two templates might seem minimal, both deserve a study since the computational cost related to the LNS might favor the simpler run and kick strategy when the running time granted to the solver is limited. We denote with the symbol ▷ the token-ring sequence and with a superscript plus ($^+$) sign the LNS.

4 Experimental Analysis

In this section, we first introduce the benchmark instance and the general settings of our analysis, and then we move to the experimental results.

4.1 Benchmark Instances

In order to accommodate the additional information that we need, we have to modify the input file format of BACP. Rather than extending it, we have decide to define a completely new format for GBACP, along the lines of the ctt format defined in [4] for the timetabling problem defined for the International Timetabling Competition (ITC-2007) track 3. This new format, explained in the website, is in our opinion more easily parsable using any programming language than the one used in CSPLib.

Six new instances, called UD1–UD6, have been extracted from the database of our university. Actually, given that the database contains historical data, many more could be created, but they would have been quite similar to these ones, therefore we decided to keep only the structurally different ones. We have also

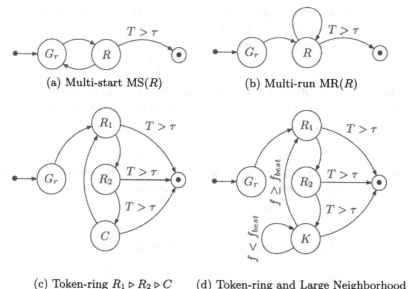

(a) Multi-start MS(R)   (b) Multi-run MR(R)

(c) Token-ring $R_1 \triangleright R_2 \triangleright C$ (d) Token-ring and Large Neighborhood search $R_1 \triangleright R_2 \triangleright (K)^+$

Fig. 2. The GLSM templates studied

Table 1. Features of the instances

Instance	Courses	Periods (Years × Terms)	Curricula	Prerequisites	Preferences
bcap8	46	8 (4 × 2)	1	33	0
bacp10	42	10 (5 × 2)	1	33	0
bacp12	66	12 (6 × 2)	1	65	0
UD1	307	9 (3 × 3)	37	1383	90
UD2	268	6 (2 × 3)	20	174	79
UD3	236	9 (3 × 3)	31	1092	66
UD4	139	6 (2 × 3)	16	188	40
UD5	282	6 (3 × 2)	31	397	54
UD6	264	4 (2 × 2)	20	70	55

translated the three CSPLib instances into our GBACP format, summing up to 9 cases to experiment on.

Table 1 summarizes the main features of the instances. All instances are available from the web, along with the format description, our best solutions, and the C++ source code of the validator that *certifies* their scores.

The format of the solution file is simply the sequence of the periods assigned to the courses listed in the same order of the input file.

4.2 General Settings and Implementation

All the algorithms have been implemented in the C++ language, exploiting the EASYLOCAL++ framework [5]. EASYLOCAL++ is a tool for local search that provides the full control structures of the algorithms, which, in its current version, also supports a limited implementation of GLSM.

The experiments were performed on an Intel QuadCore PC (64 bit) running Debian Linux 4.0, the software is compiled using the GNU C++ compiler (v. 4.1.2).

In order to compare different techniques in a fair way, we decided to grant to all of them the same total time. That is, the only condition to exit the GLSM is the timeout expired. The timeout is set to 60 seconds of CPU time on the above described PC.

The parameters of the basic components have been tuned according to the results of preliminary experiments on the single search components. The values are reported in Table 2, where c denotes the number of courses in the input instance and $N(s_0)$ is the neighborhood of solution s_0.

Table 2. Parameter settings of the algorithms

Algorithm	Parameter	Value
Dynamic Tabu Search	k_{min}	$c/4$
	k_{max}	$c/4 + 15$
	γ	1.06
Simulated Annealing	T_{start}	$\max_{s \in N(s_0)} \Delta f$
	β	0.99
	σ_N	2000

4.3 Experimental Results

We first report the results of our basic strategies on the CSPLib instances. For these "easy" instances we report only the outcome of single runs. In detail, Table 3 reports (for 1000 runs) the percentage of times an optimal (0 cost) solution has been reached and the average running time (in seconds) of all runs. The running times reported are the total one, i.e., including reading/writing files and preprocessing.

The table shows that non-trivial techniques (SA, TS, and, DTS) solve all three instances to optimality very quickly on almost all runs, whereas the others also solve them quickly but with much less success rate. A more detailed analysis of the differences of the techniques is performed on the more challenging instances.

Table 3. Results for CSPLib instances: values are success rate and average number of seconds to reach a feasible solution

Instance	SD %succ	SD time	FD %succ	FD time	RHC %succ	RHC time	SA %succ	SA time	TS %succ	TS time	DTS %succ	DTS time
bacp8	28.3	.0006	19.6	.0008	49.5	.1450	100.0	.0042	100.0	.0023	100.0	.0026
bacp10	6.2	.0009	2.2	.0012	33.3	.1906	100.0	.0429	100.0	.0046	100.0	.0060
bacp12	1.3	.0033	0.2	.0040	9.0	.2775	97.9	.1764	98.9	.0459	96.7	.08426

Regarding previous work, the best results in the literature are those in [9] which report a running time of 0.29, 0.59, and 1.09 secs to find to optimal solution for **bacp8**, **bacp10**, and **bacp12**, respectively. Our running times are somewhat better, however the absolute values are both small, and we cannot reach any conclusion on the comparison. But at least, these results confirm that local search techniques are indeed suitable to solve the BACP problem.

Moving to the analysis of the GLSM strategies described in Section 3.4, we perform a two step analysis. First we run the simple strategies equipped with all the search components to identify the most promising ones. Then we try to improve the performances of the selected search components by equipping the complex strategies with them.

Notice that not all combinations of simple search strategies and search components are meaningful. For example the multi run applied to the SD and FD algorithm is completely useless, since this algorithms stop as soon they got stuck in a local minimum.

The results are shown in Figure 3. In the plots it has been reported the normalized cost found by each algorithm on all the instances UD1–UD6, the data refers to 80 repetitions of each algorithm with different random seeds. The results of the runs are normalized with respect to the value of the best known cost of each instance, i.e., $f_{norm} = f/f_{best,i}$. Moreover we employ a semi-logarithmic scale on the y-axis, to enhance the differences between the different methods.

From the picture in Figure 3a we can see that, in general, the multi start strategy performs slightly better than the multi run ones. This can be explained by a early stagnation of the search, which prompts for a strong diversification of the search. Regardless of the search strategy employed, the best results for the simple strategies are obtained by the SA and DTS search components.

In the second step of the analysis we equip the complex strategies with the best algorithms found in the previous step, i.e., the SA and DTS search components. We test those algorithms followed by a kicker of length 2.

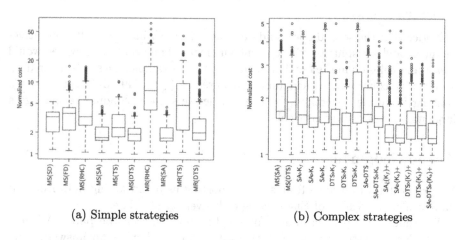

(a) Simple strategies   (b) Complex strategies

Fig. 3. Results of simple and complex strategies on instances UD1–UD6

Concerning the results for the complex strategies it is worth to notice that, in general, their results are not worse than those achieved by the simple ones (the two leftmost boxes in Fig. 3b). However, it seems that the complex strategies fits better with an intensification mechanism, e.g., DTS with both strategies or SA with the token ring followed by LNS. Moreover, the combination that gives the best performances is $SA \triangleright DTS \triangleright (K_b)^+$, which combines intensification and diversification steps in a quite complex way.

5   Conclusions and Future Work

This paper reports on an ongoing work for the solution of the GBAC problem by means of hybrid local search techniques. The preliminary results show that, although local search techniques are in general effective to solve this kind of problems, there is need to devise complex combinations of techniques to obtain better results.

In the future we plan to explore more systematically the possible GLSM combinations, in order to obtain a clearer picture of the importance of the hybridization.

Regarding the problem formulation, the GBACP proposed here seems to be adequate to solve our practical problem. Nevertheless, it still contains some limitations, that we plan to further investigate in future work. In detail, there are two main issues to address:

- Although not advisable, it is possible that a course is taken in different years in different curricula. This possibility would require a more complicated model in which courses are assigned to terms only, and for each actual pair course/curriculum we assign a different year.
- Different curricula in some cases represent free alternatives of the same degree, nevertheless some extra lower level alternatives are not modeled in the formulation. In fact, some of the curricula contain actually more courses than needed by the student to graduate, and they can drop a few of them among a larger list.

Notice that the second one could be removed simply including all alternatives in the model by splitting the current curricula. This solution however would create an explosion in the number of curricula. For example, if a student can drop 4 out of 12 courses in a curriculum, this creates $\binom{12}{8} = 495$ different curricula starting from one. Therefore, we need to devise some techniques to manage a large number of similar curricula.

Acknowledgment

We are grateful to Mauro Rainis for supporting us in extracting data from the database of the School of Engineering of University of Udine.

We thank Marco Chiarandini for helpful comments on the paper. We also thank Carlos Castro, Broderick Crawford, and Eric Monfroy for answering our questions about their work.

References

1. Ahuja, R., Ergun, Ö., Orlin, J., Punnen, A.: A survey of very-large-scale neighborhood search techniques. Discrete Applied Mathematics 123, 75–102 (2002)
2. Castro, C., Crawford, B., Monfroy, E.: A quantitative approach for the design of academic curricula. In: Smith, M.J., Salvendy, G. (eds.) HCII 2007. LNCS, vol. 4558, pp. 279–288. Springer, Heidelberg (2007)
3. Castro, C., Manzano, S.: Variable and value ordering when solving balanced academic curriculum problems. In: 6th Workshop of the ERCIM WG on Constraints (2001)
4. Di Gaspero, L., McCollum, B., Schaerf, A.: The second international timetabling competition (ITC-,: Curriculum-based course timetabling (track 3). Technical Report QUB/IEEE/Tech/ITC2007/CurriculumCTT/v1.0/1, School of Electronics, Electrical Engineering and Computer Science, Queens University, Belfast (UK). ITC-2007 (August 2007), http://www.cs.qub.ac.uk/itc2007/
5. Di Gaspero, L., Schaerf, A.: EasyLocal++: An object-oriented framework for flexible design of local search algorithms. Software—Practice and Experience 33(8), 733–765 (2003)
6. Di Gaspero, L., Schaerf, A.: Neighborhood portfolio approach for local search applied to timetabling problems. Journal of Mathematical Modeling and Algorithms 5(1), 65–89 (2006)
7. Gent, I.P., Walsh, T.: CSPLib: a benchmark library for constraints. Technical report, Technical report APES-09-1999 (1999), http://csplib.cs.strath.ac.uk/; A shorter version appears. In: The Proceedings of the 5th International Conference on Principles and Practices of Constraint Programming (CP 1999). LNCS, vol. 1713, pp. 480–481. Springer, Heidelberg (1999)
8. Hansen, P., Mladenović, N.: An introduction to variable neighbourhood search. In: Voß, S., Martello, S., Osman, I.H., Roucairol, C. (eds.) Meta-Heuristics: Advances and Trends in Local Search Paradigms for Optimization, pp. 433–458. Kluwer Academic Publishers, Dordrecht (1999)
9. Hnich, B., Kızıltan, Z., Walsh, T.: Modelling a balanced academic curriculum problem. In: CP-AI-OR 2002, pp. 121–131 (2002)
10. Hoos, H.H., Stützle, T.: Stochastic Local Search – Foundations and Applications. Morgan Kaufmann Publishers, San Francisco (2005)
11. Lambert, T., Castro, C., Monfroy, E., Saubion, F.: Solving the balanced academic curriculum problem with an hybridization of genetic algorithm and constraint propagation. In: Rutkowski, L., Tadeusiewicz, R., Zadeh, L.A., Żurada, J.M. (eds.) ICAISC 2006. LNCS (LNAI), vol. 4029, pp. 410–419. Springer, Heidelberg (2006)
12. Lourenço, H.R., Martin, O., Stützle, T.: Applying iterated local search to the permutation flow shop problem. In: Glover, F., Kochenberger, G. (eds.) Handbook of Metaheuristics. Kluwer, Dordrecht (2001)
13. Schaerf, A., Di Gaspero, L.: Measurability and reproducibility in timetabling research: Discussion and proposals. In: Burke, E., Rudová, H. (eds.) PATAT 2007. LNCS, vol. 3867, pp. 40–49. Springer, Heidelberg (2007)

Lagrangian Decomposition, Metaheuristics, and Hybrid Approaches for the Design of the Last Mile in Fiber Optic Networks

Markus Leitner[1,2] and Günther R. Raidl[2]

[1] Carinthia University of Applied Sciences
School of Telematics / Network Engineering
Klagenfurt, Austria
markus.leitner@fh-kaernten.at
[2] Vienna University of Technology
Institute for Computergraphics and Algorithms
Vienna, Austria
raidl@ads.tuwien.ac.at

Abstract. We consider a generalization of the (Price Collecting) Steiner Tree Problem on a graph with special redundancy requirements for customer nodes. The problem occurs in the design of the last mile of real-world communication networks. We formulate it as an abstract integer linear program and apply Lagrangian Decomposition to obtain relatively tight lower bounds as well as feasible solutions. Furthermore, a Variable Neighborhood Search and a GRASP approach are described, utilizing a new construction heuristic and special neighborhoods. In particular, hybrids of these methods are also studied and turn out to often perform superior. By comparison to previously published exact methods we show that our approaches are applicable to larger problem instances, while providing high quality solutions together with good lower bounds.

Keywords: Network Design, Variable Neighborhood Search, Greedy Randomized Adaptive Search Procedure, Lagrangian Relaxation, Redundancy, Steiner Tree Problem, Survivable Network Design.

1 Introduction

We consider a real-world communication network design problem arising in the expansion of existing fiber optic networks. "Fiber-to-home" has recently become economically feasible for individual households. Since the coverage of larger districts with such networks requires enormous financial resources, good algorithms for finding cost-efficient network layouts are crucial.

We consider the problem of augmenting an existing network infrastructure by additional links (and switches) in order to connect potential customer nodes. Two types of customers exist: For type-1 customers, a standard, single link connection suffices, while type-2 customers require more reliable connections,

M.J. Blesa et al. (Eds.): HM 2008, LNCS 5296, pp. 158–174, 2008.

ensuring connectivity even when a single link or routing node fails. We also consider a variant of the problem in which the redundancy condition for type-2 customers is relaxed in the sense that a connection is allowed via a final non-redundant branch that does not exceed a certain length b_{max}.

In previous work, summarized in Section 3, we approached this problem with integer linear programming techniques, including an extended multi-commodity flow formulation and a branch-and-cut algorithm. These techniques allow to find proven optimal solutions for relatively small instances.

Here we propose an approach that is also feasible for larger instances and nevertheless provides performance guarantees. It is based on a Lagrangian decomposition of the network flow model in order to obtain relatively tight lower bounds. The Lagrangian dual problem is hereby solved via the Volume Algorithm [1], which is known to often perform better than a standard subgradient search. Upper bounds and thus primal (feasible) solutions are identified at the same time, and they are improved by local search utilizing several neighborhoods. Furthermore, we propose two metaheuristic approaches based on Variable Neighborhood Search [2] and GRASP [3] to obtain primal solutions in relatively short time.

From a theoretical point-of-view, we are able to show that our Lagrangian decomposition represents a stronger model than the linear programming relaxation of the original multi-commodity network flow model. This observation is also clearly supported by our experimental results: The Volume Algorithm usually finds significantly better lower bounds in shorter times, and the obtained heuristic solutions are typically better or equal than those that could eventually be obtained by the previous approaches.

In a more general sense, this work is a good example on how Lagrangian relaxation can be applied in combination with local search based metaheuristics in order to solve a difficult practical problem heuristically and provide a quality guarantee, i.e. a lower bound, at the same time. The next section will formally introduce the problem. In Section 3 we give a short summary on related previous work. An abstract variant of the multi-commodity flow formulation from [4] is presented in Section 4 together with the Lagrangian decomposition approach for solving it. Section 5 presents the neighborhood structures used to improve solutions in the metaheuristics of Section 6 as well as in the hybrid Lagrangian approaches given in Section 7. Experimental results are discussed in Section 8, and Section 9 concludes this work.

2   Problem Definition

We are given a connected undirected graph $G = (V, E)$ representing the spatial topology of the surrounding area of potential customers. Edges in E correspond to possible cable routes and have associated lengths $l_e \geq 0$ and construction costs $c_e \geq 0$ for installing a corresponding fiber optic link. The node set V is the disjoint union of customer nodes C and spatial nodes S (switches, possible Steiner nodes). Set C is partitioned into subsets C_1 and C_2, whereby customers C_1 require a single connection (type-1) and customers C_2 need to be

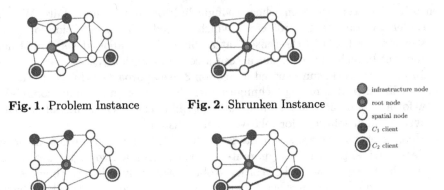

Fig. 1. Problem Instance   **Fig. 2.** Shrunken Instance

Fig. 3. Solution with $b_{max} = 0$ **Fig. 4.** Solution with $b_{max} > 0$

redundantly connected (type-2). Each customer node $k \in C$ further has associated a prize $p_k \geq 0$, i.e. expected return of investment. The already existing network infrastructure is represented by the subgraph $I = (V_I, E_I)$ of G, see Figure 1.

In a first preprocessing step, we shrink the whole existing network infrastructure, i.e. the root and all connected infrastructure nodes, into a single node $0 \in V$. From all edges connecting a node $i \in V$ to the existing infrastructure, only the cheapest edge is kept and finally replaced by an edge $(0, i)$ with the same length and costs, see Figure 2.

Let subgraph $G' = (V', E')$ with $V' \subseteq V$ and $E' \subseteq E$ represent a solution network we seek. The following conditions specify how customer nodes are to be connected:

– *Simple connection:* A customer node k from C_1 is feasibly connected iff there exists a path from node 0 to k.
– *Redundant connection:* A customer node k from C_2 is feasibly connected iff there exist two node (and edge) disjoint paths from node 0 to k, see Figure 3.
– b_{max}-*redundant connection:* Occasionally, the biconnectivity condition for the nodes in set C_2 is relaxed in the sense that such a node $k \in C_2$ may be connected to any biconnected (Steiner or customer) node $j \in V$ (the *branch-node* of k) via a single path of maximum total length $b_{max}(k) > 0$. This (optional) single path is called *branch-line* and $b_{max}(k)$ the *maximum branch-line length* for customer k, see Figure 4.

Regarding the objective, we distinguish between two alternative goals:

– In the *Operative Planning Task* (OPT) we focus on finding a minimum-cost subgraph G' feasibly connecting all customers C, with the total costs being

$$c_{OPT}(G') = \sum_{e \in E'} c_e. \tag{1}$$

This case can be considered a generalization of the classical *Steiner tree problem on a graph* (STP) where a special form of redundancy is required for the nodes in C_2.

- In the *Strategic Simulation Task* (SST) customers' prizes are also considered, and the objective is to only connect a subset $C' \subseteq C$ of customers so that the costs for building the network minus the earned prizes are minimized. In order to always have positive total costs, which eases some parts of our algorithms and notations, we perform a simple transformation by adding the constant $\sum_{k \in C} p_k$ to the objective function, yielding

$$c_{\text{SST}}(G') = \sum_{e \in E'} c_e - \sum_{k \in C'} p_k + \sum_{k \in C} p_k = \sum_{e \in E'} c_e + \sum_{k \in C \setminus C'} p_k. \qquad (2)$$

This problem variant is generalization of the *price-collecting Steiner tree problem* (PCSTP).

As already the classical Steiner tree problem on a graph is NP-hard [5], this obviously also holds for both of our problem variants. In the following presentation of our solution approaches, we primarily consider the more complex SST case if not explicitly stated and assume $p_k = \infty$, $\forall k \in C$ for the OPT case.

3   Previous Work

The Steiner Tree Problem (STP) has been considered by a lot of authors, see e.g. [6] for a survey. The Price Collecting Steiner Tree Problem (PCSTP) was introduced by Segev [7] who considered the Node Weighted STP, which is a special version of the PCSTP. The term "price collecting" has been introduced by Balas [8] for the Price Collecting Traveling Salesman Problem. A survey on methods for Survivable Network Design which can be seen as a more general version of our problem can be found in [9].

In our first attempt described in [4], we modeled this problem as an integer linear program (ILP) by means of an extended multi-commodity network flow (MCF) formulation. With the general purpose ILP-solver CPLEX [10], instances with up to 190 total nodes, 377 edges but only 6 customer nodes could be solved to proven optimality, and instances up to 2804 nodes, 3082 edges and 12 customer nodes could be solved with a final gap of about 7%. Unfortunately, this approach turned out to be unsuitable for larger instances and/or in particular instances with larger number of customer nodes, as already solving the linear programming (LP) relaxation of the ILP requires too much time due to the huge number of variables involved.

In [11], we approached this problem with a different formulation based on directed connectivity constraints. While this formulation involves only a reasonable number of variables, the number of inequalities is exponentially large. By using a branch-and-cut algorithm, however, this model could be solved relatively well, and we were able to find proven optimal solutions for instances with up to 190 nodes, 377 edges, and 13 customer nodes. For larger, practical instances

this approach unfortunately still is not applicable at all or finds quite poor solutions with huge LP-gaps only. Finally, another even stronger model based on
directed connectivity constraints which does not consider $b_{\max}$ redundancy has
been presented in [12].

4   Abstract ILP Model and Lagrangian Decomposition

To formulate this problem as an abstract ILP, we utilize decision variables $x_e \in \{0,1\}$, $\forall e \in E$, indicating whether or not edge e is part of the solution, i.e.
$x_e = 1 \leftrightarrow e \in E'$. For customer nodes $k \in C$ variables $y_k \in \{0,1\}$ denote whether
or not feasible connections according to the customers' types and $b_{\max}(k)$ exist.
Our model is based on the MCF formulation from [4], but all the different types
of flow variables for each customer $k \in C$ on directed arcs are replaced by simple
variables $f_e^k \in \{0,1\}$, $\forall e \in E$, indicating whether or not edge e is part of the
single path (type-1) or pair of disjoint paths plus the eventual branch-line (type-
2) for connecting customer k; f^k denotes the vector of all these variables for a
customer k.

Let F_k, $\forall k \in C$, be the set of all incidence vectors on E corresponding to
feasible connections for customer k. We can now formulate the SST-variant of
our problem in the following abstract way:

$$\text{minimize} \quad \sum_{e \in E} c_e x_e + \sum_{k \in C} p_k (1 - y_k) \tag{3}$$

$$\text{s.t.} \quad f_e^k \leq x_e \qquad\qquad \forall k \in C, \forall e \in E \tag{4}$$

$$f^k \in F_k \text{ if } y_k = 1 \qquad\qquad \forall k \in C \tag{5}$$

$$f_e^k \in \{0,1\} \qquad\qquad \forall k \in C, \forall e \in E \tag{6}$$

$$x_e \in \{0,1\} \qquad\qquad \forall e \in E \tag{7}$$

The objective function (3) uses variables x_e and y_k but otherwise corresponds
to (2). Inequalities (4) are called *coupling constraints* and enforce an edge to
appear in the solution when it is used for connecting at least one customer.
Conditions (5) ensure feasible connections for all selected customers ($y_k = 1$).
The OPT-variant of the model is obtained by simply ignoring the second term
in the objective function and the conditions on y_k in (5).

Note that in this form, the model is not yet a concrete ILP, as conditions (5)
are not expressed by means of linear inequalities. Ideally, we would substitute
them by a set of linear inequalities describing the convex hull $\text{conv}(F_k)$ of all
incidence vectors of feasible connections for each customer k in dependence of
variables y_k. Unfortunately, finding a (compact) set of such inequalities is not
trivial. While this can be achieved for simple (type-1) connections via a network
flow formulation, this task is quite difficult for the biconnected case involving
branch-lines (type-2).

Our MCF model from [4] represents a concrete instantiation of this abstract
model. As can be easily shown, however, it does not contain a complete description of $\text{conv}(F_k)$ but just a formulation that is valid for integer solutions. We

conclude that the MCF-formulation from [4] therefore is not as strong as an "ideal" instantiation of the abstract model.

4.1 Lagrangian Decomposition

For a general introduction to Lagrangian relaxation and decomposition see e.g. [13]. We relax the coupling constraints (4) of our abstract model in a classical Lagrangian fashion, i.e., by substituting them with corresponding penalty terms in the objective function. This yields model LR(λ):

$$\text{minimize} \sum_{e \in E} c_e x_e + \sum_{k \in C} p_k (1 - y_k) + \sum_{k \in C} \sum_{e \in E} \lambda_{k,e} \cdot (f_e^k - x_e) = \tag{8}$$

$$= \sum_{k \in C} p_k + \sum_{e \in E} \left(c_e - \sum_{k \in C} \lambda_{k,e} \right) x_e + \sum_{k \in C} \left(\sum_{e \in E} \lambda_{k,e} f_e^k - p_k y_k \right) \tag{9}$$

$$\begin{aligned} \text{s.t.} \quad & f^k \in F_k \text{ if } y_k = 1 && \forall k \in C && (10) \\ & f_e^k \in \{0, 1\} && \forall k \in C, \ \forall e \in E && (11) \\ & x_e \in \{0, 1\} && \forall e \in E && (12) \end{aligned}$$

Parameters $\lambda_{k,e} \geq 0$, $\forall k \in C$, $\forall e \in E$, are the Lagrangian multipliers, and for any feasible instantiation of them the optimal solution of LR(λ) yields a lower bound on the optimal solution value of our original abstract model [13]

For a specific selection of λ, this relaxation can be efficiently solved as it decomposes into $|C|$ independent problems of determining individual cheapest connections for each $k \in C$ on a graph whose edge costs are $\lambda_{k,e}$ (see Section 4.2). A node k is finally connected ($y_k = 1$) and the variables f_e^k corresponding to the identified connection are set to one iff the connection pays off, i.e. $\sum_{e \in E} \lambda_{k,e} f_e^k \leq p_k$. Otherwise, the connection is discarded by setting $y_k = 0$ and $f_e^k = 0$, $\forall e \in E$. Optimal values for variables x_e, $e \in E$, are independently determined by simple inspection, i.e. $x_e = 1$ iff $c_e < \sum_{k \in C} \lambda_{k,e}$, $\forall e \in E$.

The Lagrangian dual problem is the challenge of finding an optimal vector of Lagrange multipliers λ^* so that the lower bound obtained by $LR(\lambda^*)$ becomes as large as possible. As this maximization problem is convex and piecewise linear, subgradient algorithms are well suited for this purpose [13]. While different variants of such methods exist, the *Volume Algorithm* [1] has proven to be more effective than several alternatives on various occasions [14,15], and we therefore apply it here. Also, our preliminary comparisons indicate the superiority of this algorithm over the standard subgradient strategy as described in [13]. Due to space limitations, we unfortunately cannot describe the Volume Algorithm here.

4.2 Determining an Individual Optimal Connection

In each iteration of the Volume Algorithm, we need to determine for each customer $k \in C$ the cheapest feasible connection on the graph in which each edge $e \in E$ has costs $\lambda_{k,e} \geq 0$. While a simple shortest path calculation from 0 to k

returns this connection for type-1 customers $k \in C_1$, we need to determine the cheapest pair of node-disjoint paths from 0 to k for type-2 customers $k \in C_2$. Suurballe and Tarjan [16] presented an algorithm to efficiently compute a shortest arc-disjoint pair of paths between two nodes s and t on a directed graph $G_D = (V, A)$ in time $O(|A| + |V| \log |V|)$, see also [17]. Initially a shortest path tree from s as well as the shortest path P_1 from s to t are determined and the costs of each arc (i, j) are replaced with $c_{i,j} - d_j + d_i$, with d_i and d_j representing the costs of the shortest paths from s to i and j, respectively. After reversing all arcs on P_1, a shortest path P_2 from s to t is determined on this new (residual) graph using these adapted arc costs. Finally, the cheapest arc-disjoint pair of paths is given by $P_1 \triangle P_2$.

We apply this algorithm on the *split graph* of the original graph to compute node-disjoint paths. The split graph is obtained by replacing each node $v \in V$ by a pair of nodes v', v'' connected by an arc (v', v'') with zero costs. Furthermore, for each (undirected) edge $(u, v) \in E$ arcs (u, v') and (v'', u) with the same costs (and lengths) are created. Since each node v' has only one outgoing arc and each node v'' has only one ingoing arc, any pair of edge-disjoint paths is also node-disjoint.

A simple extension of the algorithm above for the case of $b_{\max}$-redundancy (see Fig. 3) is to determine the overall cheapest combination of a shortest pair of paths to a node in the $b_{\max}$-neighborhood of a customer k and a simple path from this node to k. Although we believe that a more efficient algorithm, at least with respect to average time complexity, can be found, we currently use this extension, which increases worst time complexity by a factor proportional to the size of the $b_{\max}$ neighborhood. Unnecessary calculations can be avoided by only considering possible branch-nodes j for which $d_j < \frac{1}{2} c_{\text{curr}}$ with c_{curr} being the costs of the so far cheapest connection.

4.3   Theoretical Comparison of LR(λ^*) and the MCF Formulation

For each concrete instantiation of λ, all subproblems obtained by the Lagrangian decomposition are always solved to optimality and integrality. Therefore, $f^k \in \text{conv}(F_k)$ holds for all $k \in C$, and the abstract constraints (5) of our model (3) to (7) can be regarded as "ideally instantiated". Assuming we would be able to identify an optimal Lagrange vector λ^*, the lower bound obtained by LR(λ^*)) is at least as good as the lower bound determined by an LP-relaxation of the model. As already argued before, the MCF-formulation from [4] is weaker than an "ideal" instantiation of the abstract model. We therefore conclude that LR(λ^*)) is stronger than the LP-relaxation of the MCF-formulation. Our experimental results in Section 8 also clearly support this fact.

5   Neighborhoods for Improving Primal Solutions

Our algorithms make use of three types of neighborhoods. While the first two aim at reducing the cost of a given solution, the last type consisting of two

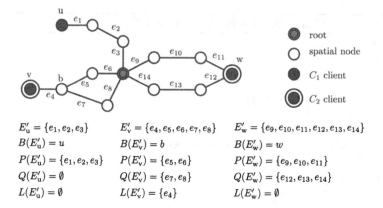

$E'_u = \{e_1, e_2, e_3\}$ $E'_v = \{e_4, e_5, e_6, e_7, e_8\}$ $E'_w = \{e_9, e_{10}, e_{11}, e_{12}, e_{13}, e_{14}\}$

$B(E'_u) = u$ $B(E'_v) = b$ $B(E'_w) = w$

$P(E'_u) = \{e_1, e_2, e_3\}$ $P(E'_v) = \{e_5, e_6\}$ $P(E'_w) = \{e_9, e_{10}, e_{11}\}$

$Q(E'_u) = \emptyset$ $Q(E'_v) = \{e_7, e_8\}$ $Q(E'_w) = \{e_{12}, e_{13}, e_{14}\}$

$L(E'_u) = \emptyset$ $L(E'_v) = \{e_4\}$ $L(E'_w) = \emptyset$

Fig. 5. An exemplary candidate solution and the representation of its connections

concrete neighborhoods tries to improve a solution by removing customers from a candidate solution. Therefore, the latter is only applicable to the SST variant.

For our neighborhood structures, a candidate solution $S' = (V', E', C', X')$ is represented, by its node set V', total edge set E', feasible connected customers $C' = \{k \in C \mid y_k = 1\}$ and individual connections $X' = \{E'_k \mid k \in C'\}$ with $E'_k = \{e \in E' \mid f^k_e = 1\}$. In other words, E'_k is the set of edges used to eventually connect customer k. Note that there may exist multiple connections to a single customer node in a solution S' in which case we store only one of them.

Furthermore, for each connection E'_k we maintain its internal structure consisting of its branch-node $B(E'_k) \in V'$, edge sets $P(E'_k), Q(E'_k) \subseteq E'$ of its two paths between 0 and $B(E'_k)$ and finally the edge set of its branch-line $L(E'_k) \subseteq E'$. Note that we assume $B(E'_k) = k$ if $b_{\max}(k) = 0$ or $k \in C_1$ as well as internally define $P(E'_k)$ to be the "first" path of a connection, i.e. $P(E'_k)$ is used for type-1 customers while $Q(E'_k) = L(E'_k) = \emptyset$ for type-1 customers, see Figure 5. Finally, to allow for efficient updates of a solution with respect to connections, we maintain for each edge $e \in E'$ a list of the customers that are connected via this edge: $M_e = \{k \in C' \mid e \in E'_k\}$.

5.1 Connection Exchange Neighborhood

The *Connection Exchange Neighborhood* (CEN) consists of all solutions differing from the current solution S' by exactly one connection E'_k, see Algorithm 1. To determine the best neighboring solution for a fixed customer $k \in C'$, CEN calculates the saving due to removing the corresponding connection E'_k (which is the sum of all edge costs exclusively used to connect k). The connection to k leading to minimum additional costs is then determined by calculating the cheapest feasible connection to k in a graph with edge costs $c'_e = 0, \forall e \in E'' = E' \setminus \{e \in E'_k \mid M_e = \{k\}\}$ and $c'_e = c_e, \forall e \in E \setminus E''$. For type-1 nodes and type-2 nodes with $b_{\max}(k) = 0$, the computational complexity of finding this new connection for one specific client node k is bounded by $O(|E| + |V| \log |V|)$. For type-2 customers with $b_{\max}(k) > 0$, we iteratively consider each possible

---

Algorithm 1. Connection Exchange (Solution S')

$c'_e = 0 \; \forall e \in E'$
$c'_e = c_e \; \forall e \in e \setminus E'$
$d_{\text{opt}} = 0$
forall $k \in C'$ **do**
$\quad E'' = \{e \in E'_k \mid M_e = \{k\}\}$
$\quad c'_e = c_e, \; \forall e \in E''$
$\quad d = \sum_{e \in E''} c_e$
$\quad E''_k = $ shortest connection to k using edge costs c'
$\quad d = \sum_{e \in E''} c_e - \sum_{e \in E''_k} c'_e$
$\quad$ **if** $d > d_{\text{opt}}$ **then**
$\quad\quad d_{\text{opt}} = d - \sum_{e \in E''_k} c'_e$
$\quad\quad$ store solution S' with E''_k replacing E'_k as best solution
$\quad c'_e = 0, \; \forall e \in E''$
return best solution

---

branch-nodes, yielding an upper bound of $O(b(|E|+|V|\log|V|))$, with b denoting the maximum number of possible branch-nodes. Therefore, the whole CEN which consists of exponentially many feasible connections can be efficiently searched for the best neighbor in $O(|C|\,b\,(|E| + |V|\log|V|))$.

5.2 Key-Path Exchange Neighborhood

A *key-node* of a solution S' is a node $v \in V' \setminus C'$ with node degree $\deg_{S'}(v) \geq 3$, while a *key-path* is a path $K_P = (V_P, E_P)$ whose end nodes are either key-nodes or customer nodes $k \in C$, while all other nodes are Steiner nodes $v \in V' \setminus (C \cup 0)$ of degree two, i.e. $\deg_{S'}(v) = 2$. This concept of key-paths is well known for the STP and several metaheuristic methods utilizing a key-path exchange neighborhood have been proposed, see e.g. [18]. The *Key-Path Exchange Neighborhood* (KPEN) given in Algorithm 2 extends this concept by exchanging key-paths while respecting node- as well as $b_{\max}$ redundancy. KPEN of a candidate solution S' consists of all feasible solutions that differ from S' by at most one key-path. To ensure feasibility, after exchanging a key-path K_P, three relevant cases need to be considered. If K_P is used to connect type-1 customer only, it may simply be replaced by any other path, while if it is used in a branch-line $L(E'_k)$ of a type-2 customer $k \in C_2$, the maximum length of the new path may be at most $b_{\max}(k) - \sum_{e \in (L(E'_k) \setminus E_P)} l_e$. Finally, if K_P is used in the first path $P(E'_k)$ of a C_2 customer k, all edges incident to "internal" nodes of its second path $Q(E'_k)$ may not be used by the new key-path to guarantee node redundancy (and vice versa for the alternate path $Q(E'_k)$). All other edges e of S' are treated as pseudo-infrastructure, i.e. $c'_e = 0$.

5.3 Connection Remove Neighborhood

Instead of exchanging a customer's connection as in CEN, the *Connection Remove Neighborhood* (CRN) removes the connection to a single customer node

---

Algorithm 2. Key Path Exchange (Solution S')

determine key-paths W
$d_{opt} = 0$
forall key-paths $(V_P, E_P) \in W$ **do**
    // *actual key-path connects its end nodes* m, n
    $c'_e = 0 \ \forall e \in E' \setminus E_P$
    $c'_e = c_e \ \forall e \in E_P \cup (E \setminus E')$
    choose $e \in E_P$ randomly
    $l_{max} = \infty$
    forall $k \in M'_e$ **do**
        if $e \in P(E'_k)$ **then**
            $c'_e = \infty, \ \forall e \in E$ incident to a inner node of $Q(E'_k)$
        else if $e \in Q(E'_k)$ **then**
            $c'_e = \infty, \ \forall e \in E$ incident to a inner node of $P(E'_k)$
        else if $e \in L(E'_k)$ **then**
            $l_{max} = b_{max}(k) - \sum_{e \in (L(E'_k) \setminus E_P)} l_e$
    $(V'_P, E'_P) = $ shortest path from m to n using c'_e with max. length l_{max}
    $d = \sum_{e \in E_P} c_e - \sum_{e \in E_{P'}} c'_e$
    if $d > d_{opt}$ **then**
        $d_{opt} = d$
        store solution S' with (V_P, E_P) replacing (V_P, E_P) as best solution

return best solution

$k \in C'$. CRN of a current solution S' therefore consists of all solutions S'', where exactly one customer connected in C' is not connected anymore, i.e. $C'' \subset C' \wedge |C'' \setminus C'| = 1$. As a customer's connection may consist of $O(|V|)$ edges only, CRN consisting of $|C'|$ neighboring solutions can be searched in $O(|C'||V|)$ time.

5.4 Restricted Two Connection Remove Neighborhood

CRN can be easily generalized to simultaneously remove multiple customer nodes. However, removing the connections to $l > 1$ customers at once will result in $|C|^l$ neighboring solutions and the computational effort of searching such a neighborhood would be $O(|C|^l|V|)$. We therefore concentrate on simultaneously removing pairs of customers $i, j \in C', i \neq j$ which share at least one edge exclusively used by them, i.e. $\exists e \in E' \mid M_e = \{i, j\}$. The *Restricted two Connection Remove Neighborhood* (R2CRN) can be searched in $O(|V| \min(|E'|, |C'|^2))$.

6 Metaheuristics

In this section we present metaheuristic approaches utilizing the neighborhoods explained in Section 5 to compute feasible solutions. After describing a construction heuristic in Section 6.1, we present a Variable Neighborhood Search (VNS) with embedded Variable Neighborhood Descent (VND) in Section 6.2 and – as an alternative – a GRASP/VND hybrid in Section 6.3.

6.1   Minimum Spanning Tree Augmentation Heuristic

We use a three-phase approach called *Minimum Spanning Tree Augmentation Heuristic* (MSTAH) to construct a feasible solution for a given selection of customers $C' \subseteq C$ to be connected. Initially, a Steiner tree G_T is computed using the *Minimum Spanning Tree* (MST) heuristic from [19]. This procedure determines a MST T_D on the *distance network*, which is the complete graph $D = (C', C' \times C')$ with node set C' and edge costs $d(u, v)$ corresponding to the costs of the cheapest paths between any $u, v \in C'$ in G. A feasible solution S'' to the Steiner Tree Problem is derived by further computing a MST on $G(T_D)$ which is the subgraph of G induced by all edges part of any cheapest path corresponding to an edge in T_D. In its second phase, MSTAH augments $S'' = (V'', E'')$ by feasible connections to C_2 customers. Such connections are determined by individually calculating the cheapest feasible connection (compare Section 4.2) for all customers $k \in C_2$. All so far selected edges $e \in E''$ are considered as pseudo-infrastructure, i.e. having zero costs. Finally, an edge minimal solution is extracted (i.e. no further edges can be deleted without violating feasibility) by greedily removing unnecessary key-paths in decreasing cost order. A similar heuristic which does not consider $b_{\max}$ redundancy has been presented in [12]. Similar to MSTAH the heuristic from [12] uses the MST heuristic [19] to compute a Steiner tree. As opposed to MSTAH redundancy for C_2 customers is ensured by adding a redundant route to each type-2 customer avoiding any inner node of the existing primary path using so far selected edge as pseudo-infrastructure.

6.2   Variable Neighborhood Search

We use the general VNS scheme with VND as embedded local improvement [2]. In VND, we alternate between CEN, KPEN, CRN, R2CRN in this order, with the latter two considered only in the SST variant.

Our shaking algorithm used to escape local optima modifies a solution S' by excluding a subset of its Steiner nodes as well as changing the set of connected customers C' in the SST variant: A set of $l = 1, \ldots, l_{\max} = |C|$ Steiner nodes $V_F \subset V' \setminus C$ of the current solution S' is randomly chosen for removal. Furthermore, we select a set of $m = \lfloor \frac{l}{3} \rfloor$ customer nodes $C_C \subset V_i' \in C$ at random. The set of customers C'' connected in the new solution S'' is $C'' = C' \triangle V_C$, i.e. we add those customers of V_C that are currently unconnected while removing the so far connected ones. Finally, we apply MSTAH using the following adapted edge costs c' with a sufficiently large value for M ($M \gg \max_{e \in E} c_e$).

$$
c'_e = \begin{cases} M & \text{if } e \text{ is incident to a nodes } v \in V_F, \\ 0 & \text{if } e \in E' \text{ and } e \text{ not incident to a node } v \in V_F, \\ c_e & \text{else.} \end{cases}
$$

Edge costs c' ensure the creation of a new solution S'' that is in general similar to S' while those Steiner nodes selected for exclusion will not be used unless there is no other option to obtain a feasible solution S''.

6.3   Greedy Randomized Adaptive Search Procedure

As an alternative to the general VNS, we also consider a GRASP in which local search is again performed by the above mentioned VND. A similar approach utilizing node- and path-based neighborhoods has been already proposed for the classical STP by Martins et al. [18]. They used a modified version of the MST heuristic [19] in the construction phase. Similarly, we modify our construction heuristic MSTAH by randomizing Kruskal's algorithm for computing the MST on the distance network D. Let $d_{\max} = \max\{d(u,v) \mid \forall(u,v) \in C' \times C'\}$ and $d_{\min} = \min\{d(u,v) \mid \forall(u,v) \in C' \times C'\}$ be the maximum and minimum distances, respectively. Instead of always adding the cheapest feasible edge that connects two yet unconnected components, the randomized spanning tree construction selects the edge to be included next randomly from a restricted candidate list consisting of all feasible edges $(u,v) \in C' \times C'$ with $d(u,v) \leq d_{\min} + \alpha(d_{\max} - d_{\min})$ with $0 < \alpha \leq 1$.

7   Combining Lagrangian Decomposition and Variable Neighborhood Descent

As described in Section 4 we solve the Lagrangian dual problem of determining optimal λ^* by the Volume Algorithm. In each iteration we need to determine optimal x_e variables as well as f_e^k variables for the current set of Lagrangian multipliers λ. The latter are computed by calculating individual cheapest connections for each customer $k \in C$ and eventually choosing to connect k in case it pays off. Obviously, the graph $S' = (V', E')$ induced by the set of edges $E' = \{e \in E \mid \exists k \text{ s.t. } f_e^k = 1\}$ is a primal feasible solution. This offers multiple ways of hybridizing the Lagrangian decomposition approach with metaheuristics in order to obtain better primal solutions and reduce the gap between lower and upper bounds.

Here, we pursue two alternatives: Either we immediately try to improve promising solutions gained by the iterations of the Volume Algorithm, or we store the N best solutions obtained by the Volume Algorithm and try to improve them after termination of the Volume Algorithm. In both cases, we use VND with CEN, KPEN, CRN, and R2CRN in this order to generate a local optimum for a given candidate solution (CRN and R2CRN are again only considered in the SST variant). According to the classification of hybrid metaheuristics given in [20] the former approach is a sequential hybridization with respect to the order of execution, while the latter falls into the category of interleaved hybridization.

As the time for performing VND on a candidate solution is not negligible, it is critical to apply it wisely on a well-chosen subset of candidate solutions only. In the interleaved approach, we found the following self-adaptive strategy with the exogenous parameters δ, γ, and $\beta_{\max}$ to work well. Let S' and S'_{best} be the current and so far best solutions obtained by the Volume algorithm, respectively. VND is applied to S' iff $c(S') \leq (1 + \beta) c(S'_{\text{best}})$. Preliminary tests indicated that a good value for β is not easy to find as it depends on the problem instance, and so we automatically adapt it each δ iterations as follows. Let r be the ratio of how

Table 1. Instance set characteristics

Set	#	$|V|$	$|E|$	$|C|$	$|\overline{C}|$	$|C_1|$	$|\overline{C_1}|$	$|C_2|$	$|\overline{C_2}|$	b_{max}	$|V(b_{max})|$
ClgSE-I1	25	190	377	5-8	5.9	3-5	3.8	2-3	2.1	30	3.79
ClgSE-I2	15	190	377	11-17	13.8	7-12	8.9	4-7	4.9	30	8.97
ClgSE-I3	15	190	377	8-12	9.6	5-8	6.0	3-6	3.6	30	6.04
ClgME-I1	25	1757	3877	6-10	7.2	4-7	5.0	2-3	2.3	100	4.96
ClgME-I2	15	1523	3290	11-14	12.2	8-11	8.7	3-4	3.5	100	8.71
ClgN1B-I1	20	2804	3082	11-14	11.8	8-11	8.5	3-4	3.3	100	8.49
ClgN1B-I2	19	2804	3082	7-11	9.0	3-6	4.1	4-6	5.0	100	3.99
ClgN1E-I1	20	3867	8477	8-14	11	3-6	4.1	5-9	6.9	150	4.12
ClgN1E-I2	20	3867	8477	10-12	10.6	6-8	6.4	4-5	4.2	150	6.39

often VND has been applied during the last δ iterations of the Volume algorithm. If $r < \gamma$ we set $\beta = \min(2\beta, \beta_{max})$ while $\beta = \max(\beta/2, \beta_{max})$ if $r > \gamma$. We chose $N = 50$, $\beta_{min} = 0.01$, $\beta_{min} = 0.4$, $\gamma = 0.05$ and $\delta = 100$ and initially set $\beta = 0.1$.

Furthermore, we memorize hash-values of candidate solution which have already been used as starting solutions to avoid unnecessary runs of VND. These hash-values are also used to ensure that the N solutions stored in the sequential approach are pairwise different.

We initialize Lagrangian multipliers by $\lambda_{k,e} = c_e/|C|$ ensuring a positive lower bound in the first iteration of the Volume Algorithm. Referring to the description of the Volume algorithm in [15], we further configured it as follows: The target value T is set to $T = 1.1z_{UB}$ with z_{UB} being the actual upper bound unless the actual lower bound $z_{LB} > 0.9\,T$ in which case T is multiplied by 1.1. We initially set $f = 0.1$ and $\alpha = 0.01$. After 20 consecutive non-improving iterations, f is multiplied by 0.67 in case it is greater than 10^{-4} and by 1.1 in an improving iteration if $f < 1$. If z_{LB} did not improve by more than 1% within the last 100 iterations and if $\alpha > 10^{-5}$, we multiply α by 0.85. The Volume Algorithm is terminated if $\lceil z_{LB} \rceil = z_{UB}$, after 250 consecutive non improving iterations, or if the maximum time limit is reached.

8   Computational Results

We used real-world instances from a German city [21] to test our approaches, see Table 1. All experiments have been performed on a single core of an Intel Xeon 5150 with 2.66GHz and 8GB RAM; ILOG CPLEX 10.0 has been used to solve the ILP for the MCF formulation from [4]. For GRASP we chose $\alpha = 0.25$ and generated 100 initial solutions, and the VNS was terminated after 100 iterations of the outermost, largest shaking move. An absolute time limit of 7200 seconds has been used for all experiments.

Table 2 compares lower bounds generated by our Lagrangian Decomposition (LD) approach to the LP-relaxation values of the MCF formulation from [4]. RED refers to the problem variant with standard redundancy constraints for C_2 customers while BMAX describes those experiments using b_{max} redundancy. As the sequential Lagrangian Decomposition approach (SEQ) as well as the interleaved approach (INT) yield similar bounds we only report the relative improvement of

Table 2. Improvement of lower bounds comp. to the LP-relaxation of MCF [4] in %.

Set	OPT+RED	SST+RED	OPT+BMAX	SST+BMAX
ClgS-I1	0.00	0.05	6.83	6.98
ClgS-I2	0.00	0.14	5.98	5.96
ClgS-I3	0.00	0.51	5.53	4.95
ClgM-I1	0.00	0.00	2.04	2.04
ClgM-I2	0.00	0.15	4.54	3.71
ClgN1B-I1	0.00	3.07	-	-
ClgN1B-I2	0.00	2.12	-	-
ClgN1E-I1	0.00	0.14	-	-
ClgN1E-I2	0.00	0.02	-	-

Table 3. Relative gaps and corresponding standard deviations in %

	Set	OPT			SST		
		LD	SEQ	INT	LD	SEQ	INT
RED	ClgS-I1	1.77 (2.45)	1.65 (2.39)	**1.63** (2.38)	1.76 (2.45)	1.65 (2.39)	**1.63** (2.38)
	ClgS-I2	12.80 (6.16)	9.12 (4.05)	**8.84** (4.08)	13.45 (7.07)	9.98 (6.18)	**9.13** (4.65)
	ClgS-I3	7.49 (6.07)	5.73 (4.81)	**5.54** (4.55)	8.89 (6.19)	7.28 (5.03)	**7.09** (4.84)
	ClgM-I1	4.29 (2.61)	2.80 (2.17)	**2.70** (2.10)	4.22 (2.62)	2.80 (2.17)	**2.61** (2.10)
	ClgM-I2	9.88 (7.10)	6.58 (4.75)	**5.89** (4.43)	11.60 (6.70)	8.50 (5.77)	**7.67** (5.65)
	ClgN1B-I1	4.12 (3.50)	2.82 (2.82)	**2.50** (2.19)	4.17 (3.45)	2.88 (2.80)	**2.58** (2.20)
	ClgN1B-I2	1.96 (1.81)	1.32 (1.43)	**1.27** (1.44)	1.84 (1.73)	1.34 (1.46)	**1.29** (1.46)
	ClgN1E-I1	3.13 (3.33)	1.51 (1.57)	**1.23** (1.24)	3.08 (3.23)	1.65 (1.81)	**1.23** (1.24)
	ClgN1E-I2	5.62 (4.67)	3.55 (2.51)	**3.21** (2.09)	5.36 (4.04)	3.53 (2.52)	**3.20** (2.08)
BMAX	ClgS-I1	2.26 (3.19)	2.13 (3.00)	**1.74** (2.40)	2.26 (3.19)	2.13 (3.00)	**1.74** (2.40)
	ClgS-I2	19.49 (7.36)	14.41 (4.46)	**12.87** (4.34)	19.53 (7.11)	14.60 (4.91)	**13.15** (4.89)
	ClgS-I3	9.05 (7.44)	6.47 (4.47)	**6.23** (4.30)	10.26 (7.67)	7.31 (4.32)	**7.14** (4.21)
	ClgM-I1	5.27 (3.22)	3.41 (2.14)	**3.09** (1.96)	5.27 (3.23)	3.34 (2.10)	**3.09** (1.96)
	ClgM-I2	15.19 (9.49)	9.29 (5.66)	**8.27** (4.53)	15.89 (9.37)	9.85 (5.86)	**9.02** (5.16)

LD in Table 2. LD generally generates equal bounds for the OPT case when b_{max}-redundancy is not considered, while the achieved lower bounds are better when dealing with the SST variant or when considering b_{max}-redundancy. The LP relaxation of the MCF formulation from [4] could not be solved for one instance of set ClgN1E-I1 (OPT variant) within 2 hours. Therefore, Table 2 reports the relative improvements for the remaining 19 instances of this set.

Table 3 compares relative gaps between upper and lower bounds generated by LD, SEQ, and INT and corresponding standard deviations (in parentheses). In general, one can observe the expected behavior that the gap increases with increasing number of customers.

SEQ and INT consistently yield for all problem variants and instances the smallest gaps, which are usually significantly better than those of LD. Table 4 depicts relative improvements of the generated upper bounds compared to LD. Without considering b_{max}-redundancy, INT generally finds solutions equally good or even better than those that could be obtained by the MCF formulation [4] within 2 hours. As the MCF formulation from [4] could not identify a feasible solution for several instances of set ClgN1E-I1 (4 instances in the OPT variant and 7 instances in the SST variant) we do not report the average improvement of MCF for this set. Average values for GRASP and VNS have been computed using 10 runs per instance.

Table 4. Relative improvement of upper bounds compared to LD in %

	Set	MCF	SEQ	INT	GRASP	VNS
OPT+RED	ClgS-I1	**0.14** (0.19)	0.12 (0.19)	**0.14** (0.19)	-0.13 (1.02)	0.12 (0.21)
	ClgS-I2	**3.40** (2.85)	3.15 (2.73)	**3.40** (2.85)	3.03 (3.23)	3.38 (2.85)
	ClgS-I3	**1.74** (2.17)	1.57 (2.16)	**1.74** (2.17)	1.48 (2.33)	1.63 (2.30)
	ClgM-I1	1.53 (1.01)	1.41 (1.12)	**1.61** (1.11)	1.54 (1.13)	1.22 (1.67)
	ClgM-I2	3.18 (2.88)	2.87 (2.74)	**3.51** (2.85)	3.23 (2.78)	2.73 (4.15)
	ClgN1B-I1	1.50 (1.82)	1.22 (1.57)	**1.51** (1.83)	1.47 (1.87)	1.41 (1.93)
	ClgN1B-I2	0.66 (1.05)	0.62 (1.02)	**0.67** (1.05)	0.53 (1.09)	**0.67** (1.05)
	ClgN1E-I1	- (-)	1.52 (1.77)	**1.78** (2.00)	1.65 (2.11)	1.14 (2.04)
	ClgN1E-I2	1.07 (3.32)	2.07 (2.41)	**2.64** (2.74)	2.56 (2.75)	2.36 (2.78)
SST+RED	ClgS-I1	**0.13** (0.19)	0.11 (0.19)	**0.13** (0.19)	-0.14 (1.02)	0.00 (0.43)
	ClgS-I2	**3.67** (2.92)	2.98 (2.90)	**3.67** (2.92)	3.30 (3.33)	2.87 (3.76)
	ClgS-I3	**1.57** (2.38)	1.40 (2.34)	**1.57** (2.38)	1.21 (2.57)	1.30 (2.43)
	ClgM-I1	1.49 (0.99)	1.35 (1.05)	**1.55** (1.04)	1.48 (1.06)	0.95 (1.89)
	ClgM-I2	3.44 (2.67)	2.71 (2.60)	**3.45** (2.62)	3.01 (2.60)	2.14 (4.14)
	ClgN1B-I1	**1.50** (1.80)	1.21 (1.51)	1.49 (1.82)	-0.86 (7.66)	0.81 (2.25)
	ClgN1B-I2	**0.54** (0.89)	0.49 (0.86)	**0.54** (0.89)	-2.68 (6.94)	-0.12 (1.83)
	ClgN1E-I1	- (-)	1.35 (1.56)	**1.75** (1.98)	1.60 (2.05)	0.36 (2.04)
	ClgN1E-I2	1.21 (2.53)	1.88 (1.95)	**2.43** (2.27)	2.10 (2.56)	1.92 (2.31)
OPT+BMAX	ClgS-I1	**0.50** (1.33)	0.12 (0.24)	0.48 (1.32)	0.23 (1.82)	0.48 (1.32)
	ClgS-I2	**5.71** (3.90)	4.08 (3.68)	5.36 (4.04)	4.97 (4.36)	5.22 (3.97)
	ClgS-I3	**2.60** (3.32)	2.18 (3.31)	2.40 (3.38)	1.68 (3.95)	2.15 (3.66)
	ClgM-I1	1.73 (2.13)	1.74 (1.69)	**2.05** (1.87)	1.84 (1.83)	1.94 (1.90)
	ClgM-I2	4.02 (6.17)	4.82 (4.99)	**5.67** (5.17)	5.51 (5.11)	5.61 (5.11)
SST+BMAX	ClgS-I1	**0.50** (1.33)	0.12 (0.24)	0.48 (1.32)	0.23 (1.82)	0.47 (1.32)
	ClgS-I2	**5.52** (4.10)	4.00 (3.23)	5.17 (4.15)	4.78 (4.41)	3.97 (4.28)
	ClgS-I3	**2.80** (3.70)	2.46 (3.78)	2.61 (3.77)	1.89 (4.26)	2.16 (4.03)
	ClgM-I1	1.62 (1.78)	1.81 (1.67)	**2.05** (1.83)	1.84 (1.83)	1.85 (1.93)
	ClgM-I2	3.59 (6.94)	4.95 (4.84)	**5.63** (5.08)	5.42 (5.09)	4.63 (4.87)

Table 5. Median run times

	Set	MCF^{LP}	MCF	LD	SEQ	INT	GRASP	VNS
OPT+RED	ClgS-*	0.2	0.9	2.0	1.7	3.7	1.7	1.2
	ClgM-*	58.2	3490.4	77.4	99.7	234.0	59.6	34.5
	ClgN1B-*	91.5	739.0	72.3	93.2	216.5	118.6	87.2
	ClgN1E-*	1103.9	7220.9	371.9	659.5	2684.9	351.6	211.9
SST+RED	ClgS-*	0.2	1.0	2.0	1.8	3.9	1.8	1.1
	ClgM-*	71.1	3052.2	90.2	109.4	226.9	58.0	30.2
	ClgN1B-*	96.1	603.5	68.2	101.1	203.2	115.3	77.2
	ClgN1E-*	824.9	7220.9	365.4	583.7	2241.2	365.7	206.8
OPT+BMAX	ClgS-*	0.3	3.2	7.7	8.4	10.5	2.0	1.5
	ClgM-*	403.6	7205.9	2865.5	3604.6	7200.0	409.8	200.5
SST+BMAX	ClgS-*	0.3	3.1	8.3	8.3	10.7	2.1	1.3
	ClgM-*	380.6	7205.9	2260.4	3401.4	6214.3	400.0	181.9

Both, GRASP and VNS also produce high quality solutions with small advantages for VNS which seem to be more stable with respect to solution quality, i.e. it almost always produces slightly better average solutions than LD. Median run times of all approaches are given in Table 5, where MCF^{LP} denotes the LP-relaxation of MCF. The CPU-times of all our approaches are in the same order of magnitude as the times for solving the LP-relaxations of the MCF formulation, but high quality feasible solutions are identified in addition to the often better lower bounds. In further tests we observed that VNS and GRASP typically produce quite good solutions very early in the search process. In that way they might be a fast alternative to solve practical instances when no performance guarantee is wanted.

9   Conclusions and Future Work

In this article we considered a generalized version of the (Price Collecting) Steiner Tree Problem where some customers have redundancy requirements. Based on an abstract version of a previously published multi-commodity flow formulation we proposed an approach based on Lagrangian decomposition which is stronger than the LP-relaxation of this MCF formulation from a theoretical point of view. Promising primal solutions are directly obtained and improved by a VND utilizing several types of neighborhoods. Furthermore, VNS and GRASP metaheuristics have been considered, making also use of the VND. Results indicate that combining Lagrangian decomposition with local search based metaheuristics produces near-optimal solutions with good performance guarantees, i.e. with relatively small gaps. In future we want accomplish a more detailed computational study with additional larger instances as well as consider an exact approach based on branch-and-price.

Acknowledgments. This work is supported by the Austrian Research Promotion Agency (FFG) under grant 811378, by the Austrian Science Fund (FWF) under grant 811378 and by the Austrian Exchange Service (Acciones Integradas, grant 13/2006). The authors further want to thank Ulrich Pferschy for fruitful discussions.

References

1. Barahona, F., Anbil, R.: The volume algorithm: producing primal solutions with a subgradient method. Mathematical Programming 87(3), 385–399 (2000)
2. Hansen, P., Mladenovic, N.: An introduction to variable neighborhood search. In: Voss, S., Martello, S., Osman, I.H., Roucairol, C. (eds.) Meta-heuristics, Advances and trends in local search paradigms for optimization, pp. 433–458. Kluwer Academic Publishers, Dordrecht (1999)
3. Feo, T., Resende, M.: Greedy randomized adaptive search procedures. Journal of Global Optimization 6(2), 109–133 (1995)
4. Wagner, D., Raidl, G.R., Pferschy, U., Mutzel, P., Bachhiesl, P.: A multi-commodity flow approach for the design of the last mile in real-world fiber optic networks. In: Waldmann, K.H., Stocker, U.M. (eds.) Operations Research Proceedings 2006, pp. 197–202. Springer, Heidelberg (2007)
5. Karp, R.M.: Reducibility among combinatorial problems. In: Miller, E., Thatcher, J.W. (eds.) Complexity of Computer Computations, pp. 85–103. Plenum Press (1972)
6. Winter, P.: Steiner problem in networks: a survey. Networks 17(2), 129–167 (1987)
7. Segev, A.: The node-weighted Steiner tree problem. Networks 17(1), 1–17 (1987)

8. Balas, E.: The prize collecting traveling salesman problem. Networks 19, 621–636 (1989)
9. Kerivin, H., Mahjoub, A.R.: Design of survivable networks: A survey. Networks 46(1), 1–21 (2005)
10. ILOG: CPLEX 10.0. (2006), http://www.ilog.com
11. Wagner, D., Pferschy, U., Mutzel, P., Raidl, G.R., Bachhiesl, P.: A directed cut model for the design of the last mile in real-world fiber optic networks. In: Fortz, B. (ed.) Proceedings of the International Network Optimization Conference 2007, Spa, Belgium, pp. 1–6, 103 (2007)
12. Chimani, M., Kandyba, M., Mutzel, P.: A new ILP formulation for 2-root-connected prize-collecting Steiner networks. In: Arge, L., Hoffmann, M., Welzl, E. (eds.) ESA 2007. LNCS, vol. 4698, pp. 681–692. Springer, Heidelberg (2007)
13. Beasley, J.E.: Lagrangean relaxation. In: Reeves, C. (ed.) Modern heuristic techniques in combinatorial problems, pp. 243–303. Blackwell Scientific Publications, Malden (1993)
14. Bahiense, L., Barahona, F., Porto, O.: Solving Steiner tree problems in graphs with Lagrangian relaxation. Journal of Combinatorial Optimization 7(3), 259–282 (2003)
15. Haouari, M., Siala, J.C.: A hybrid Lagrangian genetic algorithm for the prize collecting Steiner tree problem. Computers and Operations Research 33(5), 1274–1288 (2006)
16. Suurballe, J.W., Tarjan, R.E.: A quick method for finding shortest pairs of disjoint paths. Networks 14, 325–335 (1984)
17. Kar, K., Kodialam, M., Lakshman, T.V.: Routing restorable bandwidth guaranteed connections using maximum 2-route flows. IEEE/ACM Transactions on Networking 11(5), 772–781 (2003)
18. Martins, S.L., Resende, M.G.C., Ribeiro, C.C., Pardalos, P.M.: A parallel GRASP for the Steiner tree problem in graphs using a hybrid local search strategy. Journal of Global Optimization 17(1-4), 267–283 (2000)
19. Mehlhorn, K.: A faster approximation algorithm for the Steiner problem in graphs. Information Processing Letters 27(3), 125–128 (1988)
20. Raidl, G.R.: A unified view on hybrid metaheuristics. In: Almeida, F., et al. (eds.) HM 2006. LNCS, vol. 4030, pp. 1–12. Springer, Heidelberg (2006)
21. Bachhiesl, P.: The OPT- and the SST-problems for real world access network design – basic definitions and test instances. Working Report 01/2005, Carinthia Tech Institue, Department of Telematics and Network Engineering, Klagenfurt, Austria (2005)

Combining Forces to Reconstruct Strip Shredded Text Documents

Matthias Prandtstetter and Günther R. Raidl

Institute of Computer Graphics and Algorithms
Vienna University of Technology, Vienna, Austria
{prandtstetter,raidl}@ads.tuwien.ac.at

Abstract. In this work, we focus on the *reconstruction of strip shredded text documents* (RSSTD) which is of great interest in investigative sciences and forensics. After presenting a formal model for RSSTD, we suggest two solution approaches: On the one hand, RSSTD can be reformulated as a (standard) traveling salesman problem and solved by well-known algorithms such as the chained Lin Kernighan heuristic. On the other hand, we present a specific variable neighborhood search approach. Both methods are able to outperform a previous algorithm from literature, but nevertheless have practical limits due to the necessarily imperfect objective function. We therefore turn to a semi-automatic system which also integrates user interactions in the optimization process. Practical results of this hybrid approach are excellent; difficult instances can be quickly resolved with only few user interactions.

1   Introduction

In the fields of forensics and investigative sciences it is often required to reconstruct the information hidden on destructed paper documents. Usually, paper is destroyed by ripping up the sheets or—more professionally—by using appropriate shredding devices either producing thin strips or even small rectangles or other geometric shapes like hexagons. In this work we focus on the topic of reconstructing strip shredded text documents.

Depending on the shape, size, and the number of remnants the process of reconstructing an original document in order to restore the lost information can be very time consuming or practically almost impossible for a human. Therefore, an automatic reconstruction process is desirable. Any such approach has to acquire the strips in a first step by scanning the remnants using a (high end) scanner. Pattern recognition and image processing tasks are applied to identify the bounding boxes and orientations of the scanned strips and to gather information about features like background/paper color, text color, and other helpful features. In a second step, these attributes can be used to derive clusters of strips potentially belonging to the same original document page(s) [1]. Unfortunately any such system suffers from two drawbacks: Firstly, after the clustering process no information is directly available on how the strips have to be concatenated to form the original page(s). Secondly, any clustering approach can only marginally

M.J. Blesa et al. (Eds.): HM 2008, LNCS 5296, pp. 175–189, 2008.

reduce the problem size or even fails if many pages containing the same or similar features are shredded; examples are forms, tables, and any other regularly structured document.

Motivated by these two drawbacks, we propose a new approach to the *reconstruction of strip shredded text documents* (RSSTD) by firstly specifying the problem as a combinatorial optimization problem and secondly reformulating it as the well known *traveling salesman problem* (TSP). Furthermore, to overcome problems implied by the special structure of the resulting TSP and unavoidable inaccuracies introduced by the general modeling, a new *variable neighborhood search* (VNS) that is embedded in a system allowing user interaction is presented. Our practical results show that this approach combines and leverages machine power and human experience, knowledge, and intuition in an effective way, enabling the resolution of larger and/or more difficult RSSTD instances.

This article is structured as follows: In the next section an overview on previous and related work is given. Afterwards, our problem is formally specified. In Section 4 the transformation to the TSP is described, and Section 5 discusses possible definitions of the cost function related to the formulation as combinatorial optimization problem. Then two approaches for solving the given problem are presented—one based on the well known Lin Kernighan heuristics for the TSP and one based on a VNS and a system for integrating human interaction. Section 7 discusses results obtained by using our methods. Conclusions are drawn in Section 8.

2   Related and Previous Work

Although RSSTD is of great interest not only for intelligence agencies or forensics but also for different scientific communities, there exists not much work covering exactly this topic. A related but at the same time very different challenge is the automated solving of *jigsaw puzzles*. The major difference is the fact that for jigsaw puzzles each piece has a mostly unique shape and therefore the pure geometric information of an element can be exploited well in the reconstruction process. Furthermore and in contrast to most text documents, the image and color information on the puzzle pieces can be utilized efficiently [2].

Another related topic is the *reconstruction of manually torn paper documents*. There, shape information can also be exploited to some degree but may also be misleading due to shearing effects. The first of three major approaches was presented by Justino *et al.* [3]. They extract characteristics of the edges of snippets and then try to cling them together by iteratively matching the extracted features [3]. They state in their work that the application of the proposed method is limited to small instances of up to 15 snippets from one page.

In his master thesis, Schüller [4] proposed to use *integer linear programming* based methods for exactly reconstructing manually torn documents. The techniques presented in this work rely only on geometric information extracted from the remnants and solely focus on the borders of pages to be reconstructed since border pieces provide more reliable information and are easier to handle. Again, the application of the algorithms is limited to small instances.

De Smet [5] tries to exploit information implied by the relative order of snippets in a stack of recovered remnants. The proposed methods are limited to scenarios without missing snippets as well as a perfect snippet order. No details on how to adapt the solution process to non perfect situations are given.

In contrast to the above mentioned methods, Skeoch [6] focuses on the reconstruction of strip shredded documents but mainly discusses the scanning process and related properties of paper strips. Further, she presents a *genetic algorithm* including crossover and mutation operators as well as heuristics for generating initial solutions to restore shredded images. In contrast to text documents, a large amount of different colors usually exists in images and soft color transitions dominate. This aspect can be efficiently exploited.

Ukovich *et al.* [7] tried not to reconstruct the original document pages but to build clusters of strips belonging to the same sheet of paper by using MPEG-7 descriptors for this task. In [1], they introduced among others features like background and text color, line spacing and number of lines to be extracted from documents and discussed the potential of clustering methods.

Lately, Morandell [8] formulated the RSSTD as a combinatorial optimization problem related to the TSP. He also presents basic ideas on how to solve this new formulation by means of metaheuristics including variable neighborhood search, iterated local search, and simulated annealing. The results presented within this thesis are promising and encouraged us to pursue this approach in more detail.

3   Formal Problem Specification

In this section, we present a formal problem description of RSSTD as a combinatorial optimization problem.

We are given a finite set S of n rectangular shaped and (almost) equally sized paper snippets—so called strips—which have been produced by shredding one ore more sheet(s) of paper. In this work the widths of the strips are not further investigated since no information exploited in our approach can be extracted from them. Furthermore, the heights of all strips are assumed to be the same. If this is not the case, then a preprocessing step using clustering methods as proposed in [1] can be performed. Each set of strips having the same heights in the resulting partitioning can be used as input for our approach to RSSTD.

Although many printers are capable of duplex printing nowadays, most documents—especially in offices, one of the main application areas of shredders—are still blank on the back face. Motivated by this observation and for simplicity our presented model only regards the front face of the scanned strips. However, an extension to handle two-sided documents is possible in a straightforward way. Further, we neglect all strips of any input instance with no useful information on them. That is, all completely blank strips as well as strips with blank borders but non-empty inner regions are eliminated. Applying such a blank strip elimination procedure has two advantages. Firstly, symmetries implied by arbitrarily swapping blank strips are removed, and secondly—and more importantly—the search space is significantly reduced.

A solution $x = \langle \pi, o \rangle$ to RSSTD consists of a permutation $\pi : S \rightarrow \{1, \ldots, n\}$ of the elements in set S as well as a vector $o = \langle o_1, \ldots o_n \rangle \in \{\text{up}, \text{down}\}^n$ which assigns an orientation to each strip $s \in S$:

$$o_s = \begin{cases} \text{up} & \text{if strip } s \text{ is to be placed in its original orientation,} \\ \text{down} & \text{if strip } s \text{ is rotated by } 180°. \end{cases} \qquad (1)$$

While π_i denotes the strip at position i, $i = 1, \ldots, n$, we denote the position of a given strip $s \in S$ by $p_s \in \{1, \ldots, n\}$; i.e. $\pi_i = s \leftrightarrow p_s = i$. By $\sigma = \langle s_j, \ldots, s_k \rangle$, with $1 \leq j, k \leq n$, we denote a possibly empty (sub-)sequence of strips in a given solution. Two sequences are concatenated by the $\cdot$ operator.

In the following we make use of a cost function $c(s, s', o_s, o_{s'}) \geq 0$ to be explained later in detail, which shall provide an approximate measure for the likelihood that two strips s and s' appear side-by-side and oriented according to o_s and $o_{s'}$ in the original document, i.e. correct solution. A value of zero indicates that the contacting borders match perfectly; the larger the cost value, the more different are these borders. The overall objective is to find a solution, i.e. permutation and corresponding orientation vector, such that the following total costs are minimized:

$$\text{obj}(x) = \text{obj}_l + \sum_{i=1}^{n-1} c(\pi_i, \pi_{i+1}, o_i, o_{i+1}) + \text{obj}_r \qquad (2)$$

$$\text{obj}_l = c(\beta, \pi_1, o_\beta, o_1) \qquad (3)$$

$$\text{obj}_r = c(\pi_n, \beta, o_n, o_\beta) \qquad (4)$$

Hereby β denotes an additional (artificial) blank strip which is inserted at the beginning and the end of the page(s) to be reconstructed. This is motivated by the fact, that in most cases—especially if all strips of the original sheets of paper have been recovered—the left and right document margins are blank. As the costs of matching two blank borders are zero, omitting the additional terms obj_l and obj_r would most likely lead to a solution where the first and last strips of a correct solution are placed side-by-side. Since strip β is blank, its orientation o_β does not have any impact.

One crucial part in solving RSSTD as stated above is a proper definition of the cost function $c(s, s', o_s, o_{s'})$. A detailed discussion on this topic is given in Section 5. In any case, a cost function used for RSSTD has to have the so called *skew-symmetry property* which states that the costs for placing strip s' right to strip s have to be the same as for rotating both strips by $180°$ and placing strip s right to strip s'.

Before considering approaches for solving RSSTD, we show the following complexity result.

Theorem 1. *RSSTD is $\mathcal{NP}$-hard.*

Proof. Any (symmetric) *traveling salesman problem* (TSP) instance can be transformed into a RSSTD instance by introducing a strip for each city and defining

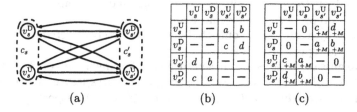

	v_s^U	v_s^D	$v_{s'}^U$	$v_{s'}^D$
v_s^U	—	—	a	b
v_s^D	—	—	c	d
$v_{s'}^U$	d	b	—	—
$v_{s'}^D$	c	a	—	—

	v_s^U	v_s^D	$v_{s'}^U$	$v_{s'}^D$
v_s^U	—	0	c_{+M}	d_{+M}
v_s^D	0	—	a_{+M}	b_{+M}
$v_{s'}^U$	c_{+M}	a_{+M}	—	0
$v_{s'}^D$	d_{+M}	b_{+M}	0	—

$$\text{(a)} \qquad\qquad \text{(b)} \qquad\qquad \text{(c)}$$

Fig. 1. In (a) a subgraph representing two strips s and s' in an AGTSP instance is depicted while in (b) the same subgraph after performing the transformation to TSP is shown. The bold lines indicate two corresponding tours.

the cost function $c(s, s', o_s, o_{s'})$ in correspondence to the TSP's distances; orientations are ignored. An arbitrary city can be chosen as RSSTD's artificial blank strip β corresponding to the left and right margins. An optimal solution to the RSSTD instance obtained in this way obviously will also correspond to an optimal solution of the original TSP. $\qquad\qquad\qquad\qquad\qquad\qquad\qquad\qquad\qquad$ □

4 Reformulation as Traveling Salesman Problem

In this section, we present a polynomial time transformation from for the RSSTD into a TSP, thus the reverse direction than in the proof above, with the motivation to find RSSTD solutions via algorithms for the TSP. To achieve this, a representation of RSSTD as an *asymmetric generalized traveling salesman problem* is developed first, and in a second step, we transform this problem into a TSP.

4.1 Formulation as Asymmetric Generalized Traveling Salesman Problem

In the *asymmetric generalized traveling salesman problem* (AGTSP) a directed graph $G = (V, A)$, with V being the set of nodes and A being the set of arcs, as well as a partitioning of V into m disjoint, non-empty clusters C_i, $i = 1, \ldots, m$, is given. Furthermore, a weight $w_a > 0$ is associated with each $a \in A$. A feasible solution to AGTSP is a tour $T \subseteq A$ that visits exactly one node of each cluster C_i while minimizing the expression $\sum_{a \in T} w_a$.

The following steps have to be performed for formulating RSSTD as AGTSP:

1. Introduce a cluster C_s for each strip $s \in S$ consisting of two vertices v_s^U and v_s^D representing the possible orientations of the corresponding strip s.
2. Introduce a cluster C_β for the virtual blank strip β and insert one vertex v_β into this cluster. Since β is blank no orientation information is necessary for this strip.
3. Each pair (s, s') of strips induces eight arcs representing the possible placements of s and s' in relation to each other, see also Fig. 1a. For instance, arc $(v_s^D, v_{s'}^U)$ represents the case that strip s' is placed right to strip s. While strip s is rotated by 180°, strip s' is positioned upright. Since strip s cannot

be placed left (or right) to itself, it is obvious that there are no arcs between two nodes representing the same strip.

4. Additionally, vertex v_β is connected via two reversely directed arcs with each other node representing a strip.

5. The weights of the arcs are chosen such that for any arc $a = (v_s^{o_s}, v_{s'}^{o_{s'}})$, with $s, s' \in \mathcal{S}$, $w_a = c(s, s', o_s, o_{s'})$. The weights for arcs leaving or entering v_β are chosen according to $c(\beta, s, o_\beta, o_s)$ or $c(s, \beta, o_s, o_\beta)$, respectively.

Obviously, an optimal solution to the AGTSP instance derived in the described way also forms a solution to the original RSSTD instance with equal costs when starting the tour at the virtual strip represented by v_β.

Several methods for solving AGTSP already exist like exact approaches, e.g. a branch-and-cut algorithm [9], as well as metaheuristics, e.g. a genetic algorithm [10]. Beside applying one of those algorithms specifically designed for solving AGTSP another possibility is to transform an AGTSP instance into a classical TSP instances and solve the latter with one of the many existing methods. In the next section we concentrate on such an approach.

4.2   Further Reformulation as TSP

The classical TSP consists of finding the shortest tour in a weighted undirected graph $G = (V, E)$ such that each vertex in V is visited exactly once. Let $w_e > 0$ be the weight associated with each edge $e \in E$. The length of a tour in TSP is computed as the sum of the tour's edge weights.

Based on the presented transformation of RSSTD to AGTSP, RSSTD can be further translated into a TSP by first applying the polynomial time transformation into a *asymmetric traveling salesman problem* (ATSP) proposed in [11] and finally applying the polynomial transformation of ATSP into TSP described in [12]. Taking a closer look at these works, two major drawbacks can be identified. On one hand, the maximum costs for edges are dramatically increased during the transformation from AGTSP into ATSP, which might lead to practical problems when trying to solve such transformed instances. On the other hand, the number of nodes in G is doubled during the translation from the asymmetric TSP to the symmetric case. Fortunately, both drawbacks can be avoided when applying a new transformation method we specifically developed for RSSTD.

Each instance of RSSTD can be transformed into an instance of TSP when first applying the reformulation as ATSP presented above and then executing the following steps. For this we adopt the idea of introducing directed cycles of zero costs within each cluster while changing the (costs of the) outgoing arcs as suggested by Behzad et al. in [11]:

1. We add two additional arcs—one in each direction—between nodes v_s^D and v_s^U for each strip $s \in \mathcal{S}$.

2. The weights of these new arcs are all set to zero.

3. In a next step, we swap the weights for $(v_s^D, v_{s'}^D)$ and $(v_s^U, v_{s'}^D)$ as well as $(v_s^D, v_{s'}^U)$ and $(v_s^U, v_{s'}^U)$. After swapping two arcs we add a constant M to the associated arc weights.

4. Since the cluster C_β consists of only one node, no transformation needs to be done for this cluster.

In Figure 1b the adjacency matrix of a subgraph of an AGTSP instance for RSSTD is presented. Figure 1c depicts the adjacency of this subgraph after applying the transformation to TSP. It can be easily checked that the resulting graph is undirected.

Theorem 2. *Any weight-minimal Hamiltonian tour on a graph obtained by the presented transformation from RSSTD can be re-transformed into an optimal placement of strips with respect to objective function* (2).

Proof. Due to the fact, that the costs for arcs connecting the nodes within a cluster are zero, any optimal tour will visit both nodes in a cluster consecutively. Assuming that there is one cluster C_i whose nodes are not visited consecutively, the tour has to enter cluster C_i at least two times. Since the costs for all arcs except for those within a cluster are equal to or greater than M, the costs of such a tour have to be greater than $(m+1) \cdot M$, with m being the number of clusters. Therefore, if M is chosen large enough, any tour, entering each cluster only once is cheaper. An appropriate value for M is $1 + m \cdot \max_{(s,s') \in S \times S} c(s, s', o_S, o'_S)$. Since each cluster is entered only once, we can decode the Hamiltonian tour as a permutation of the clusters which are representing the strips in RSSTD. Cluster C_r marks the beginning and the end of the strips' permutation. The orientation of each strip is set according to the node the cluster is entered by. If the first node visited in a cluster corresponds to the orientation *up* then the strip is oriented *up* in the corresponding solution. Analogously, orientation *down* is decoded. Further, any optimal permutation Π of strips can be transformed into an optimal tour T using the relationship described above. Assuming that there exists a tour T' with lower costs than T, we can transform T' into a permutation Π' with lower costs than Π, which is a contradiction to the assumption that Π is minimal. □

5   Definition of a Cost Function

One crucial point in RSSTD is the definition of an appropriate cost function $c(s, s', o_s, o_{s'})$ for judging the likelihood, that two strips s and s' match under their given orientations o_s and $o_{s'}$. There are several different ways on how this can be done (see also [8] on this topic), and none will be perfect in any possible situation. In this section, we discuss some important aspects on how to design a meaningful cost function for RSSTD.

As already mentioned above, any cost function for RSSTD needs to have the skew-symmetry property, i.e. placing strip s' right to strip s has to be as expensive as placing strip s right to strip s' but both rotated by 180°. To simplify the process of computing (good) lower bounds on RSSTD, we demand $c(s, s', o_s, o_{s'}) \geq 0$ always holds.

Since it is unlikely that the images of two strips with the same physical height and scanned with the same resolution significantly differ in the number of pixels

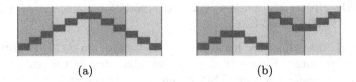

(a)   (b)

Fig. 2. Both solutions might be correct, but (a) is more likely

along the vertical edges, we assume for this work, that the number of pixels h_s along the y-axis is the same for all strips.

To simplify the next definitions, we consider eventual rotations of strips in the following as already performed; i.e. when speaking about the left side of a strip s for which $o_S = $ down, we actually refer to its original right side. The pixels on the left or right edge are those pixels which form the left or right border, respectively.

Since the majority of text documents are composed of black text on (almost) white background and we mainly focus on the reconstruction of text documents, we only consider black-and-white image data as input here. In fact, preliminary tests have shown that the usage of finer grained color or gray-scale information does not increase the quality of the solutions obtained by our approaches significantly. We remark, however, that in cases where documents contain a significant amount of different colors or gray values, an extension of our model might be meaningful and can be achieved in a more or less straightforward way.

Let $v_l(s, y, o_s), v_r(s, y, o_s) \in \{0, 1\}$ be the black-and-white values of the y-th pixel at the left and right borders of strip s under orientation o_s, respectively.

The first and most straightforward approach for defining a cost function $c_1(s, s', o_s, o_{s'})$ is by simply iterating over all pixels on the right border of strip s and compare it to the corresponding pixel on the left border of strip s'. Since we defined RSSTD as a minimization problem the value of $c_1(s, s', o_s, o_{s'})$ is increased by one if two corresponding pixels do not have the same values:

$$c_1(s, s', o_s, o_{s'}) = \sum_{y=1}^{h_s} |v_r(s, y, o_s) - v_l(s', y, o_{s'})| \qquad (5)$$

The evaluation of this cost function can be performed efficiently, but there are some situations in which it returns misleading information. For an example see the cases depicted in Figs. 2a and 2b. Of course, it is not possible to automatically decide which of the two alignments always is the correct one. Nevertheless, the situation in Fig. 2a is intuitively much more likely. Therefore, we want this alignment to receive a better evaluation than the arrangement of Fig. 2b. Hence, we adopt the idea presented in [13] to additionally consider the values of two pixels above and two pixels below to the currently evaluated position:

$$c_2(s, s', o_s, o_{s'}) = \sum_{y=3}^{h_s-2} p(s, s', o_s, o_{s'}, y) \qquad (6)$$

$$p(s, s', o_s, o_{s'}, i) = \begin{cases} 1 & \text{if } p'(s, s', o_s, o_{s'}, i) \geq \tau \\ 0 & \text{otherwise} \end{cases} \tag{7}$$

$$\begin{aligned} p'(s, s', o_s, o_{s'}, i) = |0.7 \cdot v_r(s, o_s, i) - 0.7 \cdot v_l(s', o_{s'}, i) \\ + 0.1 \cdot (v_r(s, o_s, i+1) - v_l(s', o_{s'}, i+1)) \\ + 0.1 \cdot (v_r(s, o_s, i-1) + v_l(s', o_{s'}, i-1)) \\ + 0.05 \cdot (v_r(s, o_s, i+2) + v_l(s', o_{s'}, i+2)) \\ + 0.05 \cdot (v_r(s, o_s, i-2) + v_r(s', o_{s'}, i-2))| \end{aligned} \tag{8}$$

The threshold value τ used in the definition of $p(s, s', o_s, o_{s'}, i)$ has to be chosen carefully. A good value, in particular also for handling the special case depicted in Fig. 2, is 0.1.

6 Solving RSSTD

In this section we present our concrete solution approaches for RSSTD.

6.1 Solving RSSTD Via Its Reformulation as a TSP

Using the transformation of RSSTD to TSP as presented in Section 4.2 and cost function $c_2(s, s', o_s, o_{s'})$ defined in Section 5 it is obvious to apply approaches developed for the TSP on RSSTD. Since the number of nodes in the graph underlying the TSP is always twice the number of strips in the original RSSTD instance and this number can be quite large exact algorithms might not be applicable for real world instances. Therefore, we decided to use the implementation of Applegate et al. [14] of the Chained Lin-Kernighan heuristic [15] for solving the transformed RSSTD. Detailed results are presented in Section 7.

6.2 Solving RSSTD Via VNS and Human Interaction

Even the "most precise" cost function and an exact solution of our RSSTD model will not always yield a correct arrangement fully representing the original document before destruction. The reason is that the cost function only is an (approximate) measure for the likelihood of two strips appearing next to each other. However, documents also may contain unlikely scenarios. Furthermore, text may be arranged in columns with empty parts in between. It is then impossible to find the correct order of the separated text blocks without having more specific knowledge of the documents content. Additionally applying heavier pattern recognition and knowledge extraction techniques might be feasible for certain applications but will also dramatically increase running times.

Instead, we leverage here the power of human knowledge, experience, and intuition in combination with a variable neighborhood search metaheuristic. When confronted with a candidate solution, a human often can decide quite easily which parts are most likely correctly arranged, which strips should definitely not be placed side-by-side, or which parts have a wrong orientation.

The idea of systematically integrating human interaction in an optimization process is not new. Klau *et al.* [16,17] give a survey on such approaches and present a framework called *Human Guided Search* (HuGS). The implementation is primarily based on tabu search, and the success of this human/metaheuristic integration is demonstrated on several applications.

Variable Neighborhood Search in HuGS. Since preliminary tests for solving RSSTD with tabu search as implemented in HuGS did not convince, we considered also other metaheuristics and finally decided to use a *(general) variable neighborhood search* (VNS) [18] with embedded *variable neighborhood descent* (VND) for local improvement. VNS is a metaheuristic based on the general observation that the global optimum always has to be a local optimum with respect to any possible neighborhood. The key-idea is to perform a local search and switch between multiple neighborhood structures in a well-defined way, whenever a local optimum has been reached. For more details on the general algorithm we refer to [18].

In our approach, a solution to RSSTD is represented by three arrays corresponding to the strips permutation π, the vector p storing the position for each strip, and the orientation vector o. Note that π and p are redundant, but the evaluation of the neighborhoods can be more efficiently implemented when both are available.

Neighborhoods for VNS and VND. Several different move types are used within VND and VNS. The most intuitive move is called *shifting* ($\mathcal{SH}$) and simply shifts one strip by a given amount to the right or left. More formally it can be written as

$$\mathcal{SH}(\sigma_1 \cdot \langle s_i \rangle \cdot \sigma_2 \cdot \langle s_j \rangle \cdot \sigma_3, i, j) = \sigma_1 \cdot \langle s_j \rangle \cdot \langle s_i \rangle \cdot \sigma_2 \cdot \sigma_3 \qquad (9)$$

or

$$\mathcal{SH}(\sigma_1 \cdot \langle s_j \rangle \cdot \sigma_2 \cdot \langle s_i \rangle \cdot \sigma_3, i, j) = \sigma_1 \cdot \sigma_2 \cdot \langle s_i \rangle \cdot \langle s_j \rangle \cdot \sigma_3 \qquad (10)$$

with $1 \leq i, j \leq n$. In this context σ_k denotes a possibly empty subsequence of strips. A second move, called *swapping* ($\mathcal{SW}$), is defined by swapping two arbitrary elements with each other. In a formal matter, this can be written as

$$\mathcal{SW}(\sigma_1 \cdot \langle s_i \rangle \cdot \sigma_2 \cdot \langle s_j \rangle \cdot \sigma_3, i, j) = \sigma_1 \cdot \langle s_j \rangle \cdot \sigma_2 \cdot \langle s_i \rangle \cdot \sigma_3 \qquad (11)$$

with $1 \leq i < j \leq n$. Both moves, shifting and swapping, can be extended to block moves. In the latter case, called *block swapping* ($\mathcal{BS}$), this results in a move swapping two arbitrarily long, non-overlapping subsequences of strips with each other. The other block move, namely *block shifting*, is equivalent to swapping two adjacent blocks with each other. Therefore, it is not explicitly defined in our environment. A block swap move can be formally written as

$$\mathcal{BS}(\sigma_1 \cdot \langle s_i, .., s_{i+k} \rangle \cdot \sigma_2 \cdot \langle s_j, .., s_{j+k'} \rangle \cdot \sigma_3, i, j, k, k') =$$
$$\sigma_1 \cdot \langle s_j, .., s_{j+k'} \rangle \cdot \sigma_2 \cdot \langle s_i, .., s_{i+k} \rangle \cdot \sigma_3 \qquad (12)$$

Table 1. Neighborhood structures defined for VND

neighborhood structure	$\mathcal{N}_1$	$\mathcal{N}_2$	$\mathcal{N}_3$	$\mathcal{N}_4$	$\mathcal{N}_5$
move type	$\mathcal{R}$	$\mathcal{SW}$	$\mathcal{SH}$	$\mathcal{BR}$	$\mathcal{BS}$
number of candidates	$O(n)$	$O(n^2)$	$O(n^2)$	$O(n^2)$	$O(n^4)$

with $1 \leq i < i + k < j < j + k' \leq n$. In addition to this four move types related to the assignment of strips to positions, two further moves for changing the orientation of a strip or a block of strips, called *rotating* ($\mathcal{R}$) and *block rotating* ($\mathcal{BR}$) respectively, are defined. Rotating simply rotates one strip by 180°, while block rotating executed on positions i to j first rotates all strips in this interval and in a second step swaps strips at positions i and j, $i + 1$ and $j - 1$, and so on. Using incremental evaluation schemes, each presented move can be evaluated in constant time.

In our VND, the five neighborhood structures induced by our moves are considered in the order shown in Table 1, thus, sorted by their sizes. As step function *best improvement* as well as *next improvement* have been implemented. For shaking in VNS, i random swap moves, with $1 \leq i \leq 4$, are performed. As initial solution a random solution is used.

6.3 User Interactions

For the integration of user interaction into the optimization process a set of valid user moves has to be defined. All previously described move types are contained in this set of allowed user actions. Additionally, the user can

- forbid "wrong" neighborhood relations between pairs of strips;
- lock "correct" subsequences of strips, which are concatenated and in the further optimization process considered as atomic *meta-strips*;
- lock the orientation of strips.

All of these actions also can be reverted, should the user reconsider his earlier made decisions. Our extensions of the HuGS framework provide an easy and intuitive way to visualize candidate solutions, perform the mentioned user actions, or to let VNS or the Lin Kernighan based approach continue for a while.

A main advantage of integrating human power into the search procedure is in fact that with each additional lock of strips or forbidden neighborhood relation the solution space is pruned. For example, by fixing the relative order of two strips, the number of valid solutions in the search space is divided by n.

An usual approach for a semi-automatic reconstruction of strip shredded text documents would be to first execute the TSP solver to obtain a good initial solution. Then, assuming that this solution is not already perfect, either some user moves are applied or, if there is no obvious correct subsequence of strips to be concatenated or wrongly rotated strips, VNS would be executed. Afterwards, a human inspection combined with user moves is performed. The last two steps will be repeated until either no improvement can be achieved or a solution of desired quality is obtained.

7   Experimental Results

In this section we present computational results comparing both introduced objective functions c_1 and c_2 and the different approaches. All experiments were performed on a Dual Core AMD Opteron 2214 with 4GB RAM. Both the HuGS framework and our VNS approach were implemented in Java. The Concorde TSP solver implemented by Applegate[1] was used and integrated into the Java evironment by using the *Java Native Interface*. The test instances were generated by virtually shredding paper documents, i.e. by either using scanned images or images extracted from PDF-files and cutting them into a defined number of equally sized strips. We remark that a real cutting and scanning process may loose some information or introduce errors, but neglect such effects in this work.

Quality of Solutions. As we want to find out which objective function introduced before is better suited for reconstructing strip shredded text documents, we define the *quality* of a solution as the number of correctly reconstructed subsequences of strips w.r.t. the original document. Note that the length of a correctly identified subsequence, i.e. the number of its strips, has no effect on our quality measure. This is motivated by the empiric observation that the text contained on reconstructed pages up to quality five usually can be read relatively easily. For any solutions with quality values larger than six it is typically very hard or almost impossible to the read the contained text. Further, this rating method enables us to compare results obtained for different strip widths and/or number of strips for one document.

Comparison of Results. For the results shown here we used six test instances that were shredded using different numbers of strip widths. While instances p1 to p5 consist of single text pages possessing different features (p1 and p3 are composed of continuous text only, instance p2 contains an image of a table, p4 offers a listing, and p5 shows a table with horizontal and vertical lines), instance p6 is the instance presented in [1] and consists of 10 pages with both printed and handwritten text. After virtually shredding the pages, a preprocessing step is performed on all instances, such that blank strips are eliminated.

Table 2 lists results obtained by applying the TSP solver on instances p1 to p6. We solved the instances using objective function c_1 as well as objective function c_2 and limited the CPU-time to 5 and 50 seconds, respectively. All values are average qualities over 30 runs. It can be observed that—especially for instances p2, p4 and p6—the qualities obtained by using function c_2 are remarkable better than those obtained by using c_1. Even for the short runs the standard deviations are very small and the improvement on the quality is not notable if the time limit is raised to 50 seconds. Log files show that in most cases the final solution was found after 0.5 seconds. In particular for the 10-page instance p6, the results are remarkably good. For 150 strips and cost function c_2 only 3 or 4 of the 10 pages were solved to quality 2; all others have quality 1. For 300 strips only 2 pages were always solved to quality 1 but for comparison with

---

[1] Code available at www.tsp.gatech.edu/concorde/.

Table 2. Average qualities of final solutions from the TSP solver comparing cost functions c_1 and c_2. Standard deviations are given in parentheses.

page		p1		p2		p3		p4		p5		p6	
time		5 s.	50 s.	5 s.	50 s.	5 s.	50 s.	5 s.	50 s.	5 s.	50 s.	5 s.	50 s.
30 strips	c_1	1.4 (0.5)	2.0 (0.0)	2.4 (1.4)	4.0 (0.0)	1.5 (0.5)	1.0 (0.0)	1.5 (0.5)	2.0 (0.0)	1.3 (0.5)	2.0 (0.0)	1.6 (0.5)	2.0 (0.0)
	c_2	1.4 (0.5)	1.0 (0.0)	1.5 (0.5)	2.0 (0.0)	1.6 (0.5)	2.0 (0.0)	1.7 (0.5)	2.0 (0.0)	1.5 (0.5)	1.0 (0.0)	1.6 (0.5)	2.0 (0.0)
50 strips	c_1	1.6 (0.5)	2.0 (0.0)	9.4 (0.7)	9.0 (0.0)	1.6 (0.5)	1.0 (0.0)	5.4 (0.5)	5.0 (0.0)	9.4 (0.5)	10.0 (0.0)	1.3 (0.5)	2.0 (0.0)
	c_2	1.4 (0.5)	2.0 (0.0)	4.1 (0.7)	5.0 (0.0)	1.5 (0.5)	2.0 (0.0)	1.4 (0.5)	1.0 (0.0)	1.4 (0.5)	2.0 (0.0)	1.5 (0.5)	2.0 (0.0)
100 strips	c_1	4.6 (0.5)	2.0 (0.0)	18.2 (0.8)	18.0 (0.0)	1.5 (0.5)	1.0 (0.0)	20.4 (0.5)	17.0 (0.0)	15.4 (0.5)	15.0 (0.0)	1.3 (0.4)	1.4 (0.5)
	c_2	1.5 (0.5)	2.0 (0.0)	11.8 (1.2)	13.0 (0.0)	1.4 (0.5)	1.0 (0.0)	3.8 (1.6)	5.0 (0.0)	5.5 (0.5)	6.0 (0.0)	1.4 (0.5)	2.0 (0.0)
150 strips	c_1	5.5 (0.6)	7.0 (0.0)	31.9 (0.7)	34.0 (0.0)	1.5 (0.5)	2.0 (0.0)	27.2 (1.0)	29.0 (0.9)	37.7 (0.5)	34.5 (0.5)	14.8 (0.8)	4.6 (0.5)
	c_2	1.5 (0.5)	2.0 (0.0)	26.5 (0.5)	25.0 (0.0)	1.5 (0.5)	1.0 (0.0)	16.7 (0.9)	16.0 (0.0)	9.4 (0.5)	6.0 (0.0)	4.5 (0.5)	5.0 (0.0)
300 strips	c_1	38.6 (0.7)	27.6 (0.5)	108.1 (0.8)	103.3 (1.1)	7.5 (0.5)	8.0 (0.0)	67.5 (0.6)	65.3 (0.9)	93.3 (1.1)	83.8 (0.7)	107.1 (1.6)	15.7 (1.0)
	c_2	1.6 (0.5)	2.0 (0.0)	78.3 (0.6)	73.0 (0.0)	1.5 (0.5)	1.0 (0.0)	41.5 (0.5)	43.0 (0.0)	27.4 (0.5)	27.0 (0.0)	14.3 (0.7)	14.0 (0.0)

Table 3. Average qualities of final solutions when applying VNS comparing cost functions c_1 and c_2. Standard deviations are given in parentheses.

page		p1		p2		p3		p4		p5		p6	
impr		next	best	next	best	next	best	next	best	next	best	next	best
30 strips	c_1	2.0 (0.0)	2.0 (0.0)	2.8 (1.3)	3.0 (1.4)	2.0 (0.0)	2.0 (0.0)	2.0 (0.0)	2.0 (0.0)	2.0 (0.0)	2.0 (0.0)	1.0 (0.0)	1.0 (0.0)
	c_2	2.0 (0.0)	2.0 (0.0)	2.0 (0.0)	2.0 (0.0)	2.0 (0.0)	2.0 (0.0)	2.0 (0.0)	2.0 (0.0)	2.0 (0.0)	2.0 (0.0)	1.0 (0.0)	1.0 (0.0)
50 strips	c_1	4.0 (0.0)	4.0 (0.0)	11.6 (1.4)	11.6 (1.6)	2.0 (0.0)	2.0 (0.0)	4.3 (1.2)	4.7 (1.7)	10.2 (0.4)	10.1 (0.5)	1.0 (0.0)	1.0 (0.0)
	c_2	2.0 (0.0)	2.0 (0.0)	4.7 (1.2)	5.8 (2.3)	2.0 (0.0)	2.0 (0.0)	3.2 (0.4)	3.3 (0.5)	2.1 (0.4)	2.2 (0.9)	1.0 (0.0)	1.0 (0.0)
100 strips	c_1	2.5 (1.5)	3.0 (2.1)	20.5 (2.2)	20.7 (2.2)	2.1 (0.7)	2.4 (1.2)	13.2 (3.3)	14.0 (2.8)	17.8 (2.8)	19.0 (3.0)	1.0 (0.0)	1.0 (0.2)
	c_2	2.0 (0.0)	2.0 (0.0)	14.8 (2.5)	15.5 (3.1)	2.0 (0.0)	2.0 (0.0)	7.1 (1.7)	6.6 (1.8)	6.2 (0.6)	6.5 (0.9)	1.0 (0.0)	1.0 (0.0)
150 strips	c_1	27.7 (6.7)	26.8 (8.4)	37.3 (2.0)	38.9 (2.4)	25.6 (7.6)	27.8 (9.6)	27.8 (2.2)	28.7 (3.1)	41.4 (7.3)	45.6 (7.4)	4.8 (1.5)	4.9 (1.4)
	c_2	19.5 (7.1)	22.4 (6.6)	26.0 (1.8)	27.2 (1.7)	16.8 (6.8)	16.7 (9.6)	18.7 (2.5)	18.7 (1.9)	19.6 (7.6)	23.8 (9.6)	5.6 (1.4)	4.4 (0.8)

the results presented in [1] we performed also tests with 340 strips on instance p6. This time 16 out 30 runs were solved to optimality for all other only one page was solved to quality 2 while all other were completely reconstructed. Especially when considering the time limit of 5 seconds, our methods clearly outperform those from Ukovich *et al.* [1].

Average results obtained when applying VNS without human interaction are presented in Table 3. For examining the neighborhoods we tested with both *next* as well as *best improvement* strategies, and no iteration or time limit was given. Again, the values presented are from 30 runs. We used the order of neighborhoods as presented in Section 6.2 but omitted the examination of the block swapping neighborhood N_5 for instances with more than 100 strips as the size of this neighborhood is in $O(n^4)$. We can observe that the results obtained for objective c_2 are in general better than or equal to the results obtained for c_1, but no conclusions can be drawn which step function performs better for RSSTD. Based on the poorer performance of VNS on instances with more than 100 strips we conclude that neighborhood N_5 substantially contributes to the success of VNS.

Finally we tested out semi-automatic system as it would be used in practice for reconstructing strip shredded text documents. With only few user interactions we were able to quickly restore all original documents by exploiting the benefits of the hybridization of machine and human power.

8   Conclusions

In this work, we presented a polynomial time transformation of the RSSTD to the symmetric TSP. We applied a chained Lin Kernighan heuristic as well as a newly introduced VNS for solving the RSSTD and showed that both methods are competitive with each other. In particular they clearly outperform the previous method from Ukovich *et al.*

Anyway, both approaches suffer from the necessarily imperfect objective function, which is only based on estimations of the likelihoods that strips shall be placed side-by-side under given orientations. Therefore, we embedded the algorithms in the HuGS-framework and gave the user the possibility to interact with the optimization in flexible ways. This turned out to work excellently. In this semi-automatic way, all test instances could be completely restored in very short time with only few user interactions. We consider the reconstruction of strip shredded text documents therefore as a superior example, where neither metaheuristics (and other other automated optimization techniques) nor human are able to produce satisfactory results, but a hybrid approach performs very well due to the combination of the different strengths.

References

1. Ukovich, A., Zacchigna, A., Ramponi, G., Schoier, G.: Using clustering for document reconstruction. In: Dougherty, E.R., et al. (eds.) Image Processing: Algorithms and Systems, Neural Networks, and Machine Learning. Proceedings of SPIE., International Society for Optical Engineering, vol. 6064, pp. 168–179 (2006)
2. Chung, M.G., Fleck, M., Forsyth, D.: Jigsaw puzzle solver using shape and color. In: Fourth International Conference on Signal Processing 1998, ICSP 1998, vol. 2, pp. 877–880 (1998)
3. Justino, E., Oliveira, L.S., Freitas, C.: Reconstructing shredded documents through feature matching. Forensic Science International 160(2–3), 140–147 (2006)

4. Schüller, P.: Reconstructing borders of manually torn paper scheets using integer linear programming. Master's thesis, Vienna Univ. of Technology, Austria (2008)
5. De Smet, P.: Reconstruction of ripped-up documents using fragment stack analysis procedures. Forensic science international 176(2), 124–136 (2008)
6. Skeoch, A.: An Investigation into Automated Shredded Document Reconstruction using Heuristic Search Algorithms. PhD thesis, University of Bath, UK (2006)
7. Ukovich, A., Ramponi, G., Doulaverakis, H., Kompatsiaris, Y., Strintzis, M.: Shredded document reconstruction using MPEG-7 standard descriptors. In: Proceedings of the Fourth IEEE International Symposium on Signal Processing and Information Technology, 2004, pp. 334–337 (2004)
8. Morandell, W.: Evaluation and reconstruction of strip-shredded text documents. Master's thesis, Vienna University of Technology, Austria (2008)
9. Fischetti, M., González, J.J.S., Toth, P.: A branch-and-cut algorithm for the symmetric generalized traveling salesman problem. Operations Research 45, 378–394 (1997)
10. Silberholz, J., Golden, B.: The generalized traveling salesman problem: A new genetic algorithm approach. In: Baker, E.K., et al. (eds.) Extending the Horizons: Advances in Computing, Optimization, and Decision Technologies. Operations Research/Computer Science Interfaces, vol. 37, pp. 165–181. Springer, Heidelberg (2007)
11. Behzad, A., Modarres, M.: A new efficient transformation of the generalized traveling salesman problem into traveling salesman problem. In: Proceedings of the 15th International Conference of Systems Engineering, pp. 6–8 (2002)
12. Kumar, R., Haomin, L.: On asymmetric TSP: Transformation to symmetric TSP and performance bound (submitted, 1994); Journal of Operations Research
13. Balme, J.: Reconstruction of shredded documents in the absence of shape information. Working paper, Dept.of Computer Science, Yale University, USA (2007)
14. Applegate, D., Bixby, R., Chvátal, V., Cook, W.: Finding tours in the TSP. Technical Report Number 99885, Research Institute for Discrete Mathematics, Universität Bonn (1999)
15. Applegate, D., Cook, W., Rohe, A.: Chained lin-kernighan for large traveling salesman problems. INFORMS Journal on Computing 15(1), 82–92 (2003)
16. Klau, G.W., Lesh, N., Marks, J., Mitzenmacher, M., Schafer, G.T.: The HuGS platform: A toolkit for interactive optimization. In: Proc. Advanced Visual Interfaces, AVI, pp. 324–330. ACM Press, New York (2002)
17. Klau, G.W., Lesh, N., Marks, J., Mitzenmacher, M.: Human-guided search: Survey and recent results. Technical Report TR2003-07, Mitsubishi Electric Research Laboratories, Cambridge, MA, USA (2003); Submitted to Journal of Heuristics
18. Hansen, P., Mladenović, N.: Variable neighborhood search. In: Glover, F., Kochenberger, G. (eds.) Handbook of Metaheuristics, pp. 145–184. Kluwer Academic Publishers, Dordrecht (2003)

A Memetic Algorithm for the Tool Switching Problem

Jhon Edgar Amaya[1], Carlos Cotta[2], and Antonio J. Fernández[2]

[1] Universidad Nacional Experimental del Táchira (UNET)
Laboratorio de Computación de Alto Rendimiento (LCAR), San Cristóbal, Venezuela
jedgar@unet.edu.ve
[2] Dept. Lenguajes y Ciencias de la Computación, ETSI Informática,
University of Málaga, Campus de Teatinos, 29071 - Málaga, Spain
{ccottap,afdez}@lcc.uma.es

Abstract. This paper deals with the Tool Switching Problem (ToSP), a well-known problem in operations research. The ToSP involves determining a job sequence and the tools to be loaded on a machine with the goal of minimizing the total number of tool switches. This problem has been tackled by a number of algorithmic approaches in recent years. Here, we propose a memetic algorithm that combines a problem-specific permutational genetic algorithm with a hill-climbing procedure. It is shown that this combined approach outperforms each of the individual algorithms, as well as an ad-hoc beam search heuristic defined in the literature for this problem.

1 Introduction

For some time now, the manufacturing industry is more and more often demanding flexible manufacturing systems (FMSs) as an alternative to traditional rigid production systems. This increasing interest is motivated by the fact that FMSs have the ability for self-adjustment to generate different products and/or change the order of product generation, i.e., they incorporate versatility and efficiency in mass production [1]. Basically, a FMS consists of a single machine that has several slots into which different tools can be loaded. Each slot just admits one tool, and each job executed on that machine requires a particular set of tools to be done. Jobs are sequentially executed, and therefore each time a job is to be processed, the corresponding tools must be loaded in the machine magazine. Since the number of available slots is limited, it may be required at some point to perform a tool switch, i.e., removing a tool from the magazine and inserting another one in its place. In this context, tool management is a challenging task that directly influences the efficiency of flexible manufacturing systems.

Although the order of tools in the magazine is often irrelevant, the need of performing a tool switching does depend on the order in which the jobs are executed. The Tool Switching Problem (ToSP) consists of finding an appropriate job sequence in which jobs will be executed, and an associated sequence of tool switches that minimizes the number of tool loading/unloading operations in the

M.J. Blesa et al. (Eds.): HM 2008, LNCS 5296, pp. 190–202, 2008.

magazine. Clearly, this problem is specially interesting when the time needed to change a tool is a significant part of the processing time of all the jobs (and hence the tool switching policy will significantly affect the performance of the system). Different examples of the problem can be found in diverse areas such as electronic industry, metalworking industry, computer memory management, aeronautics, and manufacturing companies in general [1,2,3,4,5]. The ToSP has also a number of variants; see for example [6,7,8].

Despite the ToSP has been tackled via different optimization techniques (including exact and metaheuristics methods – see next section), to the best of our knowledge, no population-based algorithm (let alone a hybrid population-local approach) has been applied to its resolution. This paper gives a first step in this direction and demonstrates empirically that hybrid evolutionary techniques are effective solving strategies for this problem.

2   The Tool Switching Problem

Before getting to the algorithmic approaches considered for tackling the ToSP, let us describe more formally the problem, and previous related work in the literature.

2.1   Problem Formulation

As mentioned before, the ToSP involves scheduling a number of jobs on a single machine such that the resulting number of tool switches required is kept to a minimum. This can be formalized as follows: let a ToSP instance be represented by a 4-tuple, $I = (C, n, m, A)$ where

- C denotes the magazine capacity (i.e., number of available slots),
- n is the number of jobs to be processed,
- m is the total number of tools required to process all jobs (it is assumed that $C < m$; otherwise the problem is trivial).
- A is a $m \times n$ Boolean matrix termed the *incident matrix*. This matrix defines the tool requirements to execute each job, i.e., A_{ij} =TRUE if, and only if, tool i is required to execute job j.

The solution to such an instance is a sequence $J_1, \cdots, J_n$ determining the order in which the jobs are executed, and a sequence $T_1, \cdots, T_n$ of tool configurations ($T_i \subset \{1, \cdots, m\}$) determining which tools are loaded in the magazine at a certain time.

Let $\mathbb{N}_k = \{1, \cdots, k\}$ henceforth. An integer linear programming (ILP) formulation for the ToSP is shown below, using two sets of zero-one decision variables:

- $x_{jk} = 1$ if job $j \in \mathbb{N}_n$ is assigned to position $k \in \mathbb{N}_n$ in the sequence, and 0 otherwise – see Eqs. (2) and (3),
- $y_{ik} = 1$ if tool $i \in \mathbb{N}_m$ is in the magazine at time $k \in \mathbb{N}_n$, and 0 otherwise – see Eq. (4).

Processing each job requires a particular collection of tools loaded in the magazine. It is assumed that no job requires a number of tools higher than the magazine capacity, i.e., $\sum_{i=1}^{m} \delta_{A_{ij},\text{TRUE}} \leqslant C$ for all j, where δ_{ij} is Kronecker's delta. Tool requirements are reflected in Eq. (5). Following [1], we assume the initial condition $y_{i0} = 1$ for all $i \in \mathbb{N}_m$. This initial condition amounts to the fact that the initial loading of the magazine is not considered as part of the cost of the solution (in fact, no actual switching is required for this initial load). The objective function $F(\cdot)$ counts the number of switches that have to be done for a particular job sequence – see Eq. (1).

$$\min \ F(y) = \sum_{j=1}^{n} \sum_{i=1}^{m} y_{ij}(1 - y_{i,j-1}) \tag{1}$$

$$\forall j \in \mathbb{N}_n : \ \sum_{k=1}^{n} x_{jk} = 1 \tag{2}$$

$$\forall j \in \mathbb{N}_n : \ \sum_{k=1}^{n} x_{kj} = 1 \tag{3}$$

$$\forall k \in \mathbb{N}_n : \ \sum_{i=1}^{m} y_{ik} \leqslant C \tag{4}$$

$$\forall j, k \in \mathbb{N}_n \ \forall i \in \mathbb{N}_m : \ A_{ij}x_{jk} \leqslant y_{ik} \tag{5}$$

$$\forall j, k \in \mathbb{N}_n \ \forall i \in \mathbb{N}_m : \ x_{jk}, y_{ij} \in \{0,1\} \tag{6}$$

It must be noted that the general definition above can be augmented if additional constraints are posed on tools or on the magazine. For example, it might be the case that different tools require slots of different sizes (or more than one slot). This is the so-called non-uniform ToSP [9]. Be as it may, we will consider in the following the uniform ToSP as previously defined.

2.2   Related Work

References to the ToSP can be found in the literature as early as in the 60's [2]. Since then, the uniform ToSP has been tackled via many different techniques. The late 80's contributed specially to solve the problem [10,11]. Tang and Denardo [3] proposed an ILP formulation of the problem, and later Bard [1] described a non-linear integer programming formulation with a dual-based relaxation heuristic.

Heuristics-based constructive methods have also been applied to the problem. For instance Djellab *at al.* [12] tackled ToSP via a hypergraph representation and proposed a particular heuristic oriented towards minimizing the number of (total weighted) gaps in edge-projection where a projection is basically a permutation satisfying some specific constraints; the hypergraph is used to represent the relation among jobs and the needed tools. Also, Hertz *et al.* [13] described three constructive methods (i.e., FI, GENI and GENIUS) in which at each step both a job to be inserted in current tour and the best position in the tour are selected. Additionally, nearest neighbor (NN) and 2-opt search were also considered.

Exact methods have been also applied to the problem. For instance, Laporte *et al.* [14] propose two exact algorithms: a branch-and-bound approach and a a linear programming-based branch-and-cut algorithm. Precisely this last one is based on a new ILP formulation having a better linear relaxation than that proposed previously by Tang and Denardo [3]. It must be noted that these exact methods are inherently limited, since Oerlemans [15] and Crama *et al.* [16] proved formally that the ToSP is NP-hard for $C > 2$. This limitation was already highlighted in [14], where Laporte *et al.* reported that their algorithm was capable of managing instances with 9 jobs but it presented a very low success ratio for instances over 10 jobs.

Clustering/grouping methods have also been attended. For instance, Salonen et al. [17] attacked the uniform ToSP of printed circuit boards (PCBs) and described an algorithm that iterated the process of first determining a good (or even optimal) grouping of the PCBs for further sequencing them. A hierarchical job grouping technique, based on the Jaccard similarity coefficient as clustering criterion, is additionally employed to avoid identical groupings.

The use of metaheuristics has been also recently considered. So, several tabu search approaches [17,18,19] have been used in the literature. A different, and very interesting, approach has been described by Zhou *et al.* [20] that proposed a beam search algorithm. This method was demonstrated to be specially efficient and practical compared to other techniques previously presented. The reason provided to justify this efficiency was that the performance of the algorithm can be adjusted by changing the search width and the evaluation functions. We will return later to this approach since, due to its proved efficiency, it has been included in our experimental comparison.

In any case, to the best of our knowledge, no population-based algorithm has been proposed so far to solve this problem.

3   Solving the ToSP

The ToSP can be divided into three subproblems [21]: the first subproblem is *machine loading* and consists of determining the sequence of jobs; the second subproblem is *tool loading*, consisting of determining which tool to switch (if a switch is needed) before processing a job; finally, the third subproblem is *slot loading*, and consists of deciding where (i.e., in which slot) to place each tool. We are considering the uniform ToSP, and therefore only two subproblems have to be taken into account: machine loading and tool loading.

As it will be shown in next subsection, the tool loading subproblem can be optimally solved if the sequence of jobs is known by following a specific tool switching policy (described in Section 3.1) that guarantees to obtain the optimal number of tool switches for a given job sequence. Therefore, the metaheuristic effort is concentrated on the machine loading stage. For this purpose, we will consider the use of memetic algorithms (MAs). As already mentioned, the beam search heuristic defined by Zhou *et al.* [20] is used for comparison purposes in the experimental section (see Section 5) and will be also described for the sake of completeness in Section 3.2.

3.1   The KTNS Method for Tool Loading

In the context of the uniform ToSP, the cost of switching a tool is considered a constant (the same for all tools). Under this assumption, if the job sequence is fixed, the optimal tool switching policy can be determined in polynomial time using a greedy procedure termed *Keep Tool Needed Soonest* (KTNS) [1,3][1]. The functioning of this procedure is as follows:

1. At any instant, insert all the tools that are required for the current job.
2. If one or more tools are to be inserted and there are no vacant slots on the magazine, keep the tools that are needed soonest. Let $J = \langle J_1, \cdots, J_n \rangle$ be the job sequence, and let $T_i \subset \mathbb{N}_m$ be the tool configuration at time i. Let $\Xi_{ik}(J) = \min \{ j \mid (j > k) \wedge A_{i,J_j} \}$, that is, the next instant after time k at which tool i will be needed again given sequence J. If a tool has to be removed, the tool i^* maximizing $\Xi_{ik}(J)$ is chosen, i.e., remove the tools whose next usage is farther in time.

The KTNS policy states that when tool changes are necessary, the tools required for an upcoming job should be kept in the magazine. As a side remark, the tool loading problem is NP-hard in the non-uniform ToSP, even if the job sequence is known and unit loading/unloading costs are assumed [9].

3.2   A Beam Search Heuristic

The beam search algorithm defined by Zhou *et al.* [20] is a powerful approach to tackle the ToSP. Beam search is a derivative of branch-and-bound that uses a breadth-first traversal of the search tree, and incorporates a heuristic choice to keep at each level only the best (according to some *quality* measure) β nodes (the so-called *beam width*). This sacrifices completeness, but provides a very effective heuristic search approach.

The best β nodes are selected by one-step priority evaluation functions which estimate the cost of expanding the current solution. Note that nodes in the beam represent partial solutions (i.e., sequences of λ jobs $\langle J_1, \cdots, J_\lambda \rangle$ with $\lambda < n$; if $\lambda = n$ they actually represent solutions). For each node in the current level, a decision about which job will be added to the partial sequence is done. Let $\tau_j = \{ i \mid A_{ij} = \text{TRUE} \}$, i.e., the set of tools required by job j. Two simple functions are used to ensure the quality of the solutions obtained:

$$h_1(J, k) = \#(\tau_{J_k} \cap \tau_{J_{k+1}}) \tag{7}$$

$$h_2(J, k) = \#(\tau_{J_k} \cup \tau_{J_{k+1}}) \tag{8}$$

---

[1] As Błażewicz and Finke [7] point out, the KTNS property was already known to Belady[2].

where $\#S$ is the cardinality of set S. Thus, $h_1(J,k)$ returns the number of common tools needed to process job J_k and candidate job J_{k+1} to be added to the partial job sequence. As to $h_2(J,k)$, it computes the total number of tools required to process job J_k and the candidate job J_{k+1}. These functions are used to select the beam nodes in each level, trying to maximize h_1 and using h_2 (to be minimized) to break ties.

4   A Memetic Approach to the ToSP

According to the previous discussion, the role of the MA is to determine the best job sequence, such that the total number of switches is minimized. Therefore, a permutational encoding arises as the natural way to represent solutions. Next sections will be devoted to describe our evolutionary algorithm, the neighborhood structures defined on the permutational encoding, and how these are used within the evolutionary algorithm to produce a memetic algorithm.

4.1   A Population-Based Attack to the ToSP

We have considered a steady-state genetic algorithm (GA) to evolve promising job sequences: a single solution is generated in each generation, and inserted in the population replacing the worst individual. Selection is done by binary tournament. For recombination we initially explored two schemes: the well-known order crossover (OX), and a crossover scheme named *Alternating Position* (APX) that consists in select genes alternating of each parents [22]. Preliminary experiments that we executed showed that the employment of APX provided better results, in terms of solution quality, than using OX so that we elected APX as the crossover operator.

For the purposes of mutation we have considered the *block* neighborhood. This neighborhood is proposed for the ToSP in [19] and is based on swapping whole segments of contiguous positions. The resulting mutation operator is called Random Block Insertion (RBI) and works as follows:

1. A block length $b_l \in \mathbb{N}_{n/2}$ is uniformly selected at random.
2. The starting point of the block $b_s \in \mathbb{N}_{n-2b_l}$ is subsequently selected at random.
3. Finally, an insertion point b_i is selected, such that $b_s + b_l \leqslant b_i \leqslant n - b_l$, and the segments $\langle b_s, b_s + b_l \rangle$ and $\langle b_i, b_i + b_l \rangle$ are swapped.

Obviously, if the block length $b_l = 1$ then the operation reduces to a simple position swap, but this is not typically the case when performing mutation.

4.2   Local Search

A specific local search approach considered in this work is based on the well-known all-pairs neighborhood, i.e., two permutations are neighbors if they just differ in two positions of the sequence. A steepest-ascent hill climbing (HC) approach is defined on the basis of this neighborhood structure: the neighborhood $\mathcal{N}(x)$ of the current solution x is partially traversed, and the best solution found is taken as the new current solution, provided it is better than the current one (otherwise, it is considered that there is a stagnation).

Note that the exploration of the whole neighborhood is not executed as this process becomes more and more costly as the number of jobs increases e.g., for 50 jobs, the number of neighbors for a given candidate is 1225. For this reason only a set $\mathcal{N}_x \subset \mathcal{N}(x)$, with $\#\mathcal{N}_x = \alpha n$, is explored in each step of the HC method. The selection of candidates in $\mathcal{N}_x$ is done randomly.

4.3 The Memetic Proposal

On the basis of the previously described GA, we have defined a memetic algorithm (MA) by endowing the GA with a local search scheme. To be precise, we have used the all-pairs hill climbing algorithm defined in Section 4.2 just after the mutation stage. This local search is performed for a number of $maxEval$ evaluations, or until it stagnates. It must be also noted that the local search is always performed on every new individual generated.

In all our proposals (i.e., HC, GA and MA) the fitness of the candidate is obtained by the value returned after applying the KTNS method to the candidate. The objective is thus minimizing this value.

5 Experimental Results

The experiments have been performed using four different algorithms: the beam search (BS) presented in [20] and described in Section 3.2, and the three algorithms proposed in preceding section, that is to say, a steady-state permutational GA, an steepest-ascent hill climbing (HC) search and a memetic algorithm. In the case of BS, five different equally-spaced values between 1 and 5 were considered for the beam width. As to the HC, the value $\alpha = 4$ was chosen for exploration of the neighborhood. Also, when HC is working alone, each time that local search is considered stagnated this is reactivated from a randomly selected candidate; when used inside the MA as an improvement operator, the HC is executed until reaching a stagnation or a maximum number of evaluations (i.e., exactly 1000; obviously there is no reactivation from a random point). As to the GA (and subsequently to the MA), an elitist generational model replacing the worst individual of the population ($popsize = 30$, $p_X = 1.0$, $p_M = 1/n$ where n is the number of jobs i.e., number of genes per individual) with binary tournament selection has been utilized; alternating position crossover (APX) is used, and mutation is done by applying the RBI operator. As to the MA, HC was always applied to each offspring generated after the mutation step (i.e., the probability of improvement was 1.0). The election of the parameter values (including the value for α) was done after an extensive phase of experimentation with many different values. The best combinations of the values were finally selected.

As far as we know, no standard data instance exists for this problem (at least publicly available) so that we have arbitrarily selected a wide set of problem

Table 1. Problem Instances considered in the experimental evaluation. The minimum and maximum of tools required for all the jobs is indicated in second and third rows respectively. Fourth row display the bibliography reference from which the problem instance was obtained.

	$4\zeta_{10}^{10}$	$4\zeta_{10}^{9}$	$6\zeta_{15}^{15}$	$6\zeta_{15}^{12}$	$6\zeta_{15}^{20}$	$8\zeta_{20}^{15}$	$8\zeta_{20}^{16}$	$10\zeta_{20}^{20}$	$24\zeta_{20}^{30}$	$24\zeta_{20}^{36}$	$30\zeta_{40}^{40}$	$10\zeta_{30}^{25}$	$15\zeta_{40}^{40}$	$15\zeta_{30}^{40}$	$20\zeta_{40}^{60}$	$25\zeta_{50}^{40}$
Min.	2	2	3	3	3	3	3	4	9	9	11	4	6	6	7	9
Max.	4	4	6	6	6	8	8	10	24	24	30	10	15	15	20	20
Source	[13] [19]	[1] [20]	[20]	[1] [20]	[13]	[19]	[1] [20]	[1] [20]	[1] [20]	[1] [20]	[20]	[19]	[13]	[19]	[13]	[19]

instances that were attacked in [1,13,19,20]; more specifically, 16 instances were chosen with values for the number of jobs, number of tools, and machine capacity ranging in [10,50], [9,60] and [4,25] respectively. Table 1 shows the different problem instances chosen for the experimental evaluation where a specific instance with n jobs, m tools and machine capacity C is labeled as $C\zeta_n^m$.

Five different datasets[2] (i.e., incident matrixes or relations among tools and jobs) were generated randomly per instance. Each dataset was generated with the restriction, already imposed in previous works such as [13], that no job is *covered* by any other job in the sense that $\forall i, j \in [1, m] \land i \neq j : \tau_i \nsubseteq \tau_j$ where $\tau_k = \{h \mid A_{hk} = \texttt{TRUE}\}$ is defined as before, i.e., the set of tools required to process job k. The reason to enforce this constraint is to avoid the simplification of the problem by preprocessing techniques as done for instance in [1] and [20].

For GA, HC and MA, the algorithms were run 10 times (per instance and dataset) and a maximum of $\varphi n |n - C|$ evaluations[3] per run (with $\varphi > 0$). Preliminary experiments on the value of φ proved that $\varphi = 80$ is an appropriate value that allows to keep an acceptable relation between solution quality and computational cost. Regarding the BS algorithm, because of its deterministic nature, just one execution per dataset (and per value of beam width) was run and the algorithm was allowed to be executed until exhaustion (i.e., until completing the search). All methods were implemented in Java language version 1.5, and all the experiments were carried out on a PC computer Toshiba with Operating System Debian 1.6.x and 1.5 GHz/512 MB RAM. Tables 2 and 3 show the obtained results grouped by problem instance.

Several considerations can be done here. For instance, HC performed better, with respect to the best solution result, in most of the cases than BS; this is specially evident in the lower instances (i.e., those with smaller values of n), although in some instances (i.e., $10\zeta_{30}^{25}$, $15\zeta_{40}^{30}$, $20\zeta_{40}^{60}$) BS returned better solutions. However, in average, HC behaves better than BS in the lower instances although it seems evident than BS performs better than HC when the number

---

[2] All datasets are available at *http://www.unet.edu.ve/~jedgar/ToSP/ToSP.htm*

[3] Observe that the number of evaluations increases with the number of jobs (that is assumed to be related directly with the problem difficulty) and decreases when the magazine capacity increases (that, in certain form, it is also inversely related to the problem difficulty).

Table 2. Results for $n \leqslant 20$ of GA, MA, HC, and BS considering several values ($1 \leq i \leq 5$) for the beam width β. Best results are marked in boldface.

		GA	MA	HC	1 β	2 β	3 β	4 β	5 β	Evaluations
$4\zeta_{10}^{10}$	Av	8.88	8.68	8.96	10	9.8	9.6	9.6	9.6	4800
	SD	1.518	1.618	1.624	2.098	1.833	2.059	2.059	2.059	
	B	**7**	**7**	**7**	8	8	7	7	7	
$6\zeta_{10}^{15}$	Av	14.1	13.7	14.04	15.2	14.8	14.8	14.8	14.8	3200
	SD	2.012	2.09	2.097	1.47	1.47	1.47	1.47	1.47	
	B	**11**	**11**	**11**	13	13	13	13	13	
$4\zeta_{10}^{9}$	Av	8	7.86	8.04	8.4	8.4	8.4	8.4	8.4	4800
	SD	0.8	0.721	0.894	0.49	0.49	0.49	0.49	0.49	
	B	**7**	**7**	**7**	8	8	8	8	8	
$6\zeta_{15}^{12}$	Av	16.48	15.5	17	18.2	17.6	17.6	17.4	17.4	10800
	SD	2.138	1.982	1.929	0.748	1.02	1.02	1.2	1.2	
	B	13	**12**	14	17	16	16	16	16	
$6\zeta_{15}^{20}$	Av	23.62	22.38	24.08	26.2	25.8	25.2	25.2	25.2	10800
	SD	2.134	1.938	2.115	2.315	2.135	1.6	1.6	1.6	
	B	**20**	**20**	21	23	23	23	23	23	
$8\zeta_{20}^{15}$	Av	23.66	22.36	25.06	27	26	25.6	25.2	25.2	19200
	SD	3.603	3.576	3.652	3.95	4.05	4.271	4.118	4.118	
	B	**17**	**17**	20	21	21	20	20	20	
$8\zeta_{20}^{16}$	Av	28.5	26.66	29.18	29.4	29.4	29.4	29.4	29.4	19200
	SD	2.202	1.986	2.16	1.625	1.625	1.625	1.625	1.625	
	B	24	**23**	25	27	27	27	27	27	
$10\zeta_{20}^{20}$	Av	31.6	29.92	33.1	34.2	33.6	33.4	33.4	33.4	16000
	SD	2.828	2.357	2.147	3.187	2.871	2.8	2.8	2.8	
	B	**26**	**26**	29	30	30	30	30	30	
$24\zeta_{20}^{30}$	mean	26.52	24.9	27.2	33	32.6	32.4	32.4	32.4	6400
	σ	3.061	3.282	3.572	4.427	4.499	4.63	4.63	4.63	
	best	22	**20**	22	28	28	28	28	28	
$24\zeta_{20}^{36}$	mean	48.16	46.54	49.6	54	54	53.8	53.8	53.4	6400
	σ	8.999	8.807	9.481	8.198	8.198	8.328	8.328	7.605	
	best	**35**	36	37	45	45	45	45	45	
$30\zeta_{20}^{40}$	mean	42.82	41.04	44.48	52.2	50.2	50.2	50.2	50.2	16000
	σ	5.183	4.539	5.1	6.242	7.026	7.026	7.026	7.026	
	best	32	**31**	35	42	39	39	39	39	

of jobs increases over 30. Also, it is evident that GA and MA provide the best results and clearly outperform both the HC and BS algorithms.

Focusing on the evolutionary proposals, notice firstly that the permutational GA provides comparable results, in terms of best solutions, to those of MA. In fact, the GA presents roughly the same performance than the MA, but becomes clearly inferior when the average is considered as the MA always provides better results in this case.

A non-parametrical statistical test –Wilcoxon ranksum– was executed on the results returned by all the executions performed by HC, GA and MA. Then a comparison between each method with the other two was executed; Tables 4, 5

Table 3. Results for $n > 20$ of GA, MA, HC, and BS described in [20] considering several values $(1 \leq i \leq 5)$ for the beam width β. Best results are marked in boldface.

		GA	MA	HC	1 β	2 β	3 β	4 β	5 β	Evaluations
	mean	69.2	64.92	74.9	73.6	70.8	70.8	70.8	70.6	48000
$10\zeta_{30}^{25}$	σ	3.4	1.573	1.9	1.02	1.47	1.47	1.47	1.497	
	best	**62**	**62**	69	72	68	68	68	68	
	mean	105.08	100.86	111	111.6	110	109.2	107.8	107.8	36000
$15\zeta_{30}^{40}$	σ	13.335	12.9	14.323	15.187	13.55	13.407	13.257	13.257	
	best	**81**	**81**	89	89	89	89	89	89	
	mean	105.28	97.96	111.5	105.2	103.2	102.8	102.8	102.4	80000
$15\zeta_{40}^{30}$	σ	8.775	7.887	8.273	8.518	9.579	9.745	9.745	9.972	
	best	88	**86**	98	96	93	93	93	93	
	mean	220.6	211.88	231.1	221.8	220	218.8	218.6	218.6	64000
$20\zeta_{40}^{60}$	σ	8.825	7.812	9.104	7.026	6.164	5.706	5.817	5.817	
	best	**200**	201	216	215	215	215	215	215	
	mean	161.82	153.36	169	167.2	164	162.8	162.8	161.8	100000
$25\zeta_{50}^{40}$	σ	13.671	13.52	13.485	12.906	12.554	12.335	12.335	12.156	
	best	141	**132**	146	152	147	147	147	147	

Table 4. Comparison: HC vs. GA/MA. Each cell displays the number of datasets in which HC is considered significantly better than GA/MA according to Wilcoxon ranksum test.

	$4\zeta_{10}^{10}$	$4\zeta_{10}^{9}$	$6\zeta_{10}^{15}$	$6\zeta_{15}^{12}$	$6\zeta_{15}^{20}$	$8\zeta_{20}^{15}$	$8\zeta_{20}^{16}$	$10\zeta_{20}^{20}$	$24\zeta_{20}^{30}$	$24\zeta_{20}^{36}$	$30\zeta_{20}^{40}$	$10\zeta_{30}^{25}$	$15\zeta_{30}^{40}$	$15\zeta_{40}^{30}$	$20\zeta_{40}^{60}$	$25\zeta_{50}^{40}$
GA	0	0	0	0	0	0	0	0	1	0	0	0	0	0	0	0
MA	0	0	0	0	0	0	0	0	1	0	0	0	0	0	0	0

Table 5. Comparison: GA vs. HC/MA. Each cell displays the number of datasets in which GA is considered significantly better than HC/MA according to Wilcoxon ranksum test.

	$4\zeta_{10}^{10}$	$4\zeta_{10}^{9}$	$6\zeta_{10}^{15}$	$6\zeta_{15}^{12}$	$6\zeta_{15}^{20}$	$8\zeta_{20}^{15}$	$8\zeta_{20}^{16}$	$10\zeta_{20}^{20}$	$24\zeta_{20}^{30}$	$24\zeta_{20}^{36}$	$30\zeta_{20}^{40}$	$10\zeta_{30}^{25}$	$15\zeta_{30}^{40}$	$15\zeta_{40}^{30}$	$20\zeta_{40}^{60}$	$25\zeta_{50}^{40}$
HC	0	0	0	0	0	3	0	2	1	2	2	4	5	5	5	4
MA	0	0	0	0	0	0	0	0	0	0	0	0	0	0	0	0

Table 6. Comparison: MA vs. HC/GA. Each cell displays the number of datasets in which MA is considered significantly better than HC/GA according to Wilcoxon ranksum test.

	$4\zeta_{10}^{10}$	$4\zeta_{10}^{9}$	$6\zeta_{10}^{15}$	$6\zeta_{15}^{12}$	$6\zeta_{15}^{20}$	$8\zeta_{20}^{15}$	$8\zeta_{20}^{16}$	$10\zeta_{20}^{20}$	$24\zeta_{20}^{30}$	$24\zeta_{20}^{36}$	$30\zeta_{20}^{40}$	$10\zeta_{30}^{25}$	$15\zeta_{30}^{40}$	$15\zeta_{40}^{30}$	$20\zeta_{40}^{60}$	$25\zeta_{50}^{40}$
HC	2	0	0	5	5	5	5	5	3	5	5	5	5	5	5	5
GA	0	0	1	2	4	2	5	4	3	3	1	4	4	5	5	4

and 6 show the results of the comparison of HC vs. GA and MA, GA vs. HC and MA, and MA vs. HC and GA respectively. Each cell in the tables indicates the number of times that the corresponding algorithm is significantly better than the other one with respect to the 5 datasets per instance. For example, a 4 appearing in Table 6 for the instance $10\zeta_{30}^{25}$ in the row of GA means that the MA behaves significantly better, according to the statistical test, than the GA in 4 of the 5 datasets that were used to solve the specific problem instance. These results corroborate our preliminar considerations and one can observe that the GA outperforms HC when the number of jobs increases and that the MA outperforms both HC and GA and also behaves in general evidently better than the GA.

6   Conclusions and Future Work

We have tackled here the tool switching problem (in its uniform version) and have proposed three methods to attack it. Two of the methods are, as far as we know, the first evolutionary approaches to handle the tool switching problem. Combining ideas from the realm of evolutionary computation and hill climbing methods, we have specifically devised an evolutionary proposal that takes the form of a memetic algorithm. An empirical evaluation was executed in order to prove the validity and performance of the proposed techniques. For comparison purposes, we have considered in the experimentation the beam search method described in [20], as this was demonstrated to be specially efficient and practical compared to another techniques previously published.

The experiments demonstrate that the three methods (i.e., a hill climbing search, a genetic algorithm and a memetic algorithm) provide encouraging results and are capable of improving the results obtained by the BS. Focusing on our proposals, the memetic algorithm outperforms both a permutational genetic algorithm and a steepest-ascent hill climbing method. A statistical test demonstrates that the MA is significantly superior to the other two proposals.

We believe that there is room for improvement. For instance, it would be interesting to prove alternative methods to HC such as tabu search or variable neighborhood search. In this case, it would be necessary to obtain a good balance between intensification and exploration in the local search component of the MA; in our proposal we have leaned towards exploration by evaluating only a part of the whole neighborhood space, and have obtained encouraging results, but perhaps a more intensive strategy can also produce valuable results. We also plan to analyze new instances and variants of the problem [6,7,8].

Acknowledgements

Second and third authors were partially supported by Spanish MCyT under contract TIN2005-08818-C04-01. Third author was also partially supported by projects TIN2007-67134 (from Spanish Ministry of Innovation and Science) and P06-TIC2250 (from Andalusia Regional Government).

References

1. Bard, J.F.: A heuristic for minimizing the number of tool switches on a flexible machine. IIE Transactions 20(4), 382–391 (1988)
2. Belady, L.: A study of replacement algorithms for virtual storage computers. IBM Systems Journal 5, 78–101 (1966)
3. Tang, C.S., Denardo, E.V.: Models arising from a flexible manufacturing machine, part I: minimization of the number of tool switches. Oper. Res. 36(5), 767–777 (1988)
4. Privault, C., Finke, G.: Modelling a tool switching problem on a single nc-machine. Journal of Intelligent Manufacturing 6(2), 87–94 (1995)
5. Shirazi, R., Frizelle, G.: Minimizing the number of tool switches on a flexible machine: an empirical study. International Journal of Production Research 39(15), 3547–3560 (2001)
6. Kashyap, A.S., Khator, S.K.: Modeling of a tool shared flexible manufacturing system. In: WSC 1994: Proceedings of the 26th conference on Winter simulation, San Diego, CA, USA, Society for Computer Simulation International, pp. 986–993 (1994)
7. Błażewicz, J., Finke, G.: Scheduling with resource management in manufacturing systems. European Journal of Operational Research 76, 1–14 (1994)
8. Hong-Bae, J., Yeong-Dae, k., Suh, S.H.W.: Heuristics for a tool provisioning problem in a flexiblemanufacturing system with an automatic tool transporter. IEEE Transactions on Robotics and Automation 15(3), 488–496 (1999)
9. Crama, Y., Moonen, L.S., Spieksma, F.C., Talloen, E.: The tool switching problem revisited. European Journal of Operational Research 182(2), 952–957 (2007)
10. ElMaraghy, H.A.: Automated tool management in flexible manufacturing. Journal of Manufacturing Systems 4(1), 1–14 (1985)
11. Kiran, A., Krason, R.: Automated tooling in a flexible manufacturing system. Industrial Engineering 20, 52–57 (1988)
12. Djellab, H., Djellab, K., Gourgand, M.: A new heuristic based on a hypergraph representation for the tool switching problem. International Journal of Production Economics 64(1-3), 165–176 (2000)
13. Hertz, A., Laporte, G., Mittaz, M., Stecke, K.: Heuristics for minimizing tool switches when scheduling part types on a flexible machine. IIE Transactions 30, 689–694 (1998)
14. Laporte, G., Salazar-González, J., Semet, F.: Exact algorithms for the job sequencing and tool switching problem. IIE Transactions 36(1), 37–45 (2004)
15. Oerlemans, A.: Production planning for flexible manufacturing systems. Ph.d. dissertation, University of Limburg, Maastricht, Limburg, Netherlands (October 1992)
16. Crama, Y., Kolen, A., Oerlemans, A., Spieksma, F.: Minimizing the number of tool switches on a flexible machine. International Journal of Flexible Manufacturing Systems 6, 33–54 (1994)
17. Salonen, K., Raduly-Baka, C., Nevalainen, O.S.: A note on the tool switching problem of a flexible machine. Computers & Industrial Engineering 50(4), 458–465 (2006)
18. Hertz, A., Widmer, M.: An improved tabu search approach for solving the job shop scheduling problem with tooling constraints. Discrete Applied Mathematics 65, 319–345 (1993)
19. Al-Fawzan, M.A., Al-Sultan, K.S.: A tabu search based algorithm for minimizing the number of tool switches on a flexible machine. Comput. Ind. Eng. 44(1), 35–47 (2003)

20. Zhou, B.H., Xil, L.F., Cao, Y.S.: A beam-search-based algorithm for the tool switching problem on a flexible machine. The International Journal of Advanced Manufacturing Technology 25(9-10), 876–882 (2005)
21. Tzur, M., Altman, A.: Minimization of tool switches for a flexible manufacturing machine with slot assignment of different tool sizes. IIE Transactions 36(2), 95–110 (2004)
22. Larrañaga, P., Kuijpers, C., Murga, R., Inza, I., Dizdarevic, S.: Genetic algorithms for the travelling salesman problem: A review of representations and operators. Articial Intelligence Review 13, 129–170 (1999)

Author Index